Springer Tracts in Modern Physics

Volume 282

Springer Tracts in Modern Physics provides comprehensive and critical reviews of topics of current interest in physics. The following fields are emphasized:

- Elementary Particle Physics
- Condensed Matter Physics
- Light Matter Interaction
- Atomic and Molecular Physics
- Complex Systems
- Fundamental Astrophysics

Suitable reviews of other fields can also be accepted. The Editors encourage prospective authors to correspond with them in advance of submitting a manuscript. For reviews of topics belonging to the above mentioned fields, they should address the responsible Editor as listed in “Contact the Editors”.

More information about this series at http://www.springer.com/series/426

Kazutaka Nakamura

Quantum Phononics

Introduction to Ultrafast Dynamics of Optical Phonons

Kazutaka Nakamura
Laboratory for Materials and Structures,
Institute of Innovative Research
Tokyo Institute of Technology
Yokohama, Japan

ISSN 0081-3869 ISSN 1615-0430 (electronic)
Springer Tracts in Modern Physics
ISBN 978-3-030-11926-3 ISBN 978-3-030-11924-9 (eBook)
https://doi.org/10.1007/978-3-030-11924-9

Library of Congress Control Number: 2019930564

This Springer imprint is published by the registered company Springer Nature Switzerland AG
The registered company address is: Gewerbestrasse 11, 6330 Cham, Switzerland

Preface

Phonon are quantums of lattice vibrations and are bosons as well as photon. The phonons are nowadays coherently generated and controlled by using an ultrashort optical pulse. Quantum phononics is a new field in condensed-matter physics and materials science, which includes the generation, detection, and the engineering and coherent control of quantum states of lattice vibrations. This book describes the fundamentals of generation, detection, and coherent control of optical phonons using ultrashort laser pulses. While they are important for heat transfer and capacity, acoustic phonons are not included in this book. The required basic knowledge of quantum mechanics, quantum optics, solid state physics, and nonlinear spectroscopy is presented in Chaps. 1–4, which were developed during a one-quarter course for graduate students in the department of materials science at Tokyo Institute of Technology. Chapters 5–7 are based on the author's recent journal papers on coherent phonons and their coherent control. I have not tried to be encyclopedic in the treatment, either in terms of complete references or a discussion of every type of experiment and theory. Reference are included for more detailed background or historical interest. Instead, I tried to explain detailed derivations for important equations, which are often difficult for experimental physicists and materials scientists.

Chapter 1 reviews the basics of quantum mechanics, state vectors, time evolution of a quantum state, and perturbation expansion, which are used throughout the text. In Chap. 2, the density operator and a double-sided Feynman diagram are introduced, which are used to describe nonlinear optical processes. Chapter 3 explains the quantum mechanics of the harmonic oscillator, which is of essential importance in field quantization and phonon dynamics. Coherent and squeezed states, which are important in quantum optics, are introduced. Chapter 4 reviews lattice vibrations and field quantization and introduces the "phonon". In Chaps. 5 and 6, we discuss coherent phonons using ultrafast optical measurements and the quantum mechanical description of the generation and detection mechanism. Section 5.3 shows squeezed phonons. In Chap. 7, we discuss the coherent control of optical phonons using an ultrashort pulse train and its selected applications.

I especially thank Prof. Yosuke Kayanuma, my co-worker for several years, for his guidance and advice on theory. I thank my colleagues and students participated in my research projects on coherent phonons, especially Prof. Masahiro Kitajima, Prof. Oleg V. Misochko, Prof. Fujio Minami, Prof. Yutaka Shikano, Dr. Yasuaki Okano, Dr. Hiroshi Takahashi, Dr. Jianbo Hu, and Dr. Katsura Norimatsu. I also thank Gordon Han Ying Li for checking English and equations in this book. Immeasurable thanks are due to my wife Mitsuko, without her constant companionship and support this book would never have been possible. My thanks also go to Dr. Claus Ascheron, Physics Editor of Springer-Verlag at Heidelberg (now retired), for his kind support.

Yokohama, Japan Kazutaka Nakamura

Contents

Chapter 1
Time Evolution of Quantum State

In this chapter, we summarize basic knowledge of quantum mechanics required for representing a phonon and its time evolution. Ket and bra vectors are used to represent a quantum state. Time evolution of the quantum state is expressed by using three ways: Schrödinger picture, Heisenberg picture, and interaction picture. Perturbation expansion of a time-dependent ket vector is explained for the interaction picture.

1.1 Description of Quantum State

In quantum mechanics, a physical state is represented by a state vector in a complex vector space [1]. Using the Dirac notation [2], a quantum state is represented by a ket vector denoted by $|\varphi\rangle$, which can be represented by a column vector with complex components. We now introduce the bra vector, which is denoted by $\langle\varphi|$, dual to the ket vector.[1] In a two-dimensional vector space, $|\varphi\rangle$ is represented by

$$|\varphi\rangle = \begin{pmatrix} C_1 \\ C_2 \end{pmatrix}, \tag{1.1}$$

where C_1 and C_2 are complex numbers. When we set another ket vector $|\varphi'\rangle$

$$|\varphi'\rangle = \begin{pmatrix} C_1' \\ C_2' \end{pmatrix}, \tag{1.2}$$

[1]We consider a dual vector space using ket and bra vector spaces.

K. Nakamura, *Quantum Phononics*, Springer Tracts in Modern Physics 282,
https://doi.org/10.1007/978-3-030-11924-9_1

the vector inner product is defined by

$$\langle\varphi'|\varphi\rangle = \begin{pmatrix} C_1'^* & C_2'^* \end{pmatrix} \begin{pmatrix} C_1 \\ C_2 \end{pmatrix} = C_1'^* C_1 + C_2'^* C_2, \tag{1.3}$$

where C_1^* and C_2^* are complex conjugates of C_1 and C_2, respectively. The inner product of their own is

$$\langle\varphi|\varphi\rangle = C_1^* C_1 + C_2^* C_2 = |C_1|^2 + |C_2|^2. \tag{1.4}$$

$\sqrt{\langle\varphi|\varphi\rangle}$ is known as the norm and corresponds to the length of the ket vector $|\varphi\rangle$. When two vectors $|\varphi\rangle$ and $|\varphi'\rangle$ are orthogonal each other, their inner product is zero: $\langle\varphi'|\varphi\rangle = 0$.

1.2 Operator

An operator $\hat{A}$ acts on a ket vector from the left side $\hat{A}|\varphi\rangle$ and the resulting product is another ket vector. When the ket vector is represented by a N-dimensional vector, the operator $\hat{A}$ corresponds to the $N \times N$ matrix. When a vector $|\varphi\rangle$ satisfies the following equation

$$\hat{A}|\varphi\rangle = a|\varphi\rangle, \tag{1.5}$$

where a is the constant, $|\varphi\rangle$ and a are called eigenvector and eigenvalue for $\hat{A}$.

1.2.1 Hermitian Operator

The Hermitian conjugate operator for a linear operator $\hat{A}$ is expressed by $\hat{A}^\dagger$. By setting the matrix elements $A_{i,j}$, those of the Hermitian conjugate operators are defined by $(\hat{A}^\dagger)_{i,j} = (A_{j,i})^*$. When the Hermitian conjugate operator $\hat{A}^\dagger$ is the same as the original operator $\hat{A}$, this operator is called Hermitian operator. The Hermitian operator has the following properties:

- Eigenvalues are real.
- Eigenvectors, belonging to the different eigenvalues, are orthogonal each other.

Thus, the operator for the observable physical quantity is an Hermitian operator. Furthermore, eigenvectors of an Hermitian operator are complete system, and the normalized eigenvectors are used as the basis set of an orthogonal system. Any sate $|\varphi\rangle$ can be expanded using all of eigenvectors $|\phi_i\rangle$ of the Hermitian operator

$$|\varphi\rangle = \sum_i C_i |\phi_i\rangle, \tag{1.6}$$

where C_i is a complex number.

1.2.2 Projection Operator

Setting an operator $\hat{P}(a) = |a\rangle\langle a|$, in which $|a\rangle$ is normalized, the operator selects the component of a ket vector ($|\varphi\rangle$) parallel to $|a\rangle$:

$$\hat{P}(a)|\varphi\rangle = |a\rangle\langle a|\varphi\rangle = \langle a|\varphi\rangle|a\rangle, \tag{1.7}$$

where $\langle a|\varphi\rangle$ is a classical number. The operator is known as the projection operator along the base ket vector $|a\rangle$ [1]. Using the normalized orthogonal system $\{|a_i\rangle\}$, the identity operator $\hat{I}$ is expressed using the projection operator

$$\hat{I} = \sum_i \hat{P}(a_i) = \sum_i |a_i\rangle\langle a_i|. \tag{1.8}$$

When we operate the identity operator on any operator $\hat{B}$, the operator does not change:

$$\hat{B} = \hat{I}\hat{B}\hat{I} = \sum_{i,j} |a_i\rangle\langle a_i|\hat{B}|a_j\rangle\langle a_j| = \sum_{i,j} \langle a_i|\hat{B}|a_j\rangle|a_i\rangle\langle a_j|. \tag{1.9}$$

Thus, any operator can be represented as linear combinations of $|a_i\rangle\langle a_j|$. In particular, the Hermitian operator $\hat{A}$ can be expressed by a linear combination of its own eigenvectors $\{|\alpha_i\rangle\}$, where $\hat{A}|\alpha_i\rangle = \alpha_i|\alpha_i\rangle$:

$$\hat{A} = \sum_i C_i|\alpha_i\rangle|a_i\rangle\langle\alpha_i| = \sum_i C_i\hat{P}(\alpha_i). \tag{1.10}$$

This expression is called spectral decomposition of the Hermitian operator.

1.3 Measurements, Observable, and Expectation Value

A dynamical variable such as position and momentum is represented by an operator that can act on a ket vector. A measurement of the dynamical variable gives one of its eigenvalues.[2] When the system $|\varphi\rangle$ is represented by a linear combination of eigenstates $|a_i\rangle$ of the dynamical variable $\hat{A}$, $|\varphi\rangle = \sum_i C_i|a_i\rangle$, before the measurement, one of eigenstates $|a_i\rangle$ is selected and its eigenvalue a_i is obtained by the measurement. We do not know which eigenstate is selected, but the probability $\mathrm{Prob}(a_i)$ that the $|a_i\rangle$ is selected is obtained by

$$\mathrm{Prob}(a_i) = |\langle a_i|\varphi\rangle|^2 = |C_i|^2. \tag{1.11}$$

---

[2]Dirac wrote in his book "In this way we see that a measurement always causes the system to jump onto an eigenstate of the dynamical variable that is being measured, the eigenvalue this eigenstate belongs to being equal to the result of the measurement" [2]. Dirac's words are also referred to in [1].

The expectation value of $\hat{A}$ taken with respect to state $|\varphi\rangle$ is defined by

$$\begin{aligned}\langle \hat{A}\rangle &\equiv \langle\varphi|\hat{A}|\varphi\rangle \\ &= \sum_{i,j}\langle\varphi|a_j\rangle\langle a_j|\hat{A}|a_i\rangle\langle a_i|\varphi\rangle = \sum_{i,j} C_j^*\langle a_j|a_i|a_i\rangle C_i \\ &= \sum_i a_i|C_i|^2 = \sum_i \text{Prob}(a_i)a_i. \end{aligned} \tag{1.12}$$

When eigenstates of a real dynamical variable form a complete set, it is called an observable. An observable is represented by a Hermitian operator because its eigenvalues are real.

1.4 Time Evolution of a Quantum System

There are three types of representation for time evolution of a quantum system (equation of motion in quantum mechanics): Schrödinger, Heisenberg, and Interaction pictures [3].

1.4.1 Schrödinger Picture

In the Schrödinger picture, time evolution of a quantum system is represented by time evolution of state vectors. When we write the state vector at time t as $|\varphi(t)\rangle$ for a closed quantum system, time evolution of this state is defined by the Schrödinger equation:

$$i\hbar\frac{\mathrm{d}}{\mathrm{d}t}|\varphi(t)\rangle = \hat{H}|\varphi(t)\rangle, \tag{1.13}$$

where $\hat{H}$ is the Hermitian operator representing the total energy of the system. $\hat{H}$ is called the Hamiltonian and is the time-independent operator for the closed system. When there is no degeneracy, the energy eigenstate $|n\rangle$ for the eigenvalue E_n satisfies

$$\hat{H}|n\rangle = E_n|n\rangle, \tag{1.14}$$

which is called time-independent Schrödinger equation. When the state $|\varphi(t)\rangle$ starts from the energy eigenstate $|n\rangle$, $|\varphi(0)\rangle = |n\rangle$, the state at time t is expressed as

$$|\varphi(t)\rangle = e^{-iE_nt/\hbar}|n\rangle. \tag{1.15}$$

Only the phase factor is different between $|\varphi(t)\rangle$ and $|\varphi(0)\rangle$. The phase factor rotates with an angular frequency $\omega_n = E_n/\hbar$. In the closed system, the probability of detecting the energy eigenstate does not change and the energy eigenstate is a steady state. Since an arbitrary state in the closed system is expanded using energy eigenstates, any initial state $|\psi(0)\rangle$ is represented by $|\psi(0)\rangle = \sum_n C_n|n\rangle$. Then the state $|\psi(t)\rangle$ at time t is expressed by

$$|\psi(t)\rangle = \sum_n C_n e^{-iE_n t/\hbar}|n\rangle = \sum_n C_n e^{-i\omega_n t}|n\rangle. \tag{1.16}$$

1.4.1.1 Time-Evolution Operator

The norm of the state vector $|\psi(t)\rangle$ is expressed by

$$\langle\psi(t)|\psi(t)\rangle = \sum_{n,n'} C_n^* C_{n'} e^{i\omega_n t} e^{-i\omega_{n'} t}\langle n|n\rangle = \langle\psi(0)|\psi(0)\rangle, \tag{1.17}$$

and is constant throughout the time evolution. Thus, the transformation from $|\psi(0)\rangle$ to $|\psi(t)\rangle$ is a unitary transformation. We consider the unitary operator $\hat{U}(t)$ for the time evolution,

$$|\psi(t)\rangle = \hat{U}(t)|\psi(0)\rangle. \tag{1.18}$$

This unitary operator is called the time-evolution operator. Using the Schrödinger equation, the time-evolution operator satisfies

$$i\hbar\frac{\mathrm{d}}{\mathrm{d}t}\hat{U}(t) = \hat{H}\hat{U}(t), \tag{1.19}$$

with an initial condition $\hat{U}(0) = \hat{I}$. If the Hamiltonian is time independent, the time-evolution operator is expressed by using an exponential form[3]

$$\hat{U}(t) = \exp\left(\frac{-i\hat{H}t}{\hbar}\right). \tag{1.20}$$

If the Hamiltonian is time dependent, the equation for $\hat{U}(t)$ is modified as

$$i\hbar\frac{\mathrm{d}}{\mathrm{d}t}\hat{U}(t) = \hat{H}(t)\hat{U}(t). \tag{1.21}$$

[3]The exponential of a linear operator $\hat{A}$ is defined by as a power series expansion: $\exp\hat{A} \equiv \hat{I} + \hat{A} + (1/2!)\hat{A}^2 + (1/3!)\hat{A}^3 \cdots$.

Integrating this equation, we get formally

$$i\hbar\int_0^t \frac{\mathrm{d}}{\mathrm{d}t_1}\hat{U}(t_1)\mathrm{d}t_1 = \int_0^t \hat{H}(t_1)\hat{U}(t_1)\mathrm{d}t_1$$

$$i\hbar\hat{U}(t) = i\hbar\hat{U}(0) + \int_0^t \hat{H}(t_1)\hat{U}(t_1)\mathrm{d}t_1. \quad (1.22)$$

If there is no change in $|\psi(t)\rangle$ from $|\psi(0)\rangle$ at $t = 0$, then $\hat{U}(0) = 1$ and we get

$$\hat{U}(t) = 1 + \frac{1}{i\hbar}\int_0^t \hat{H}(t_1)\hat{U}(t_1)\mathrm{d}t_1. \quad (1.23)$$

Inserting $\hat{U}(t)$ defined by this equation into the integration, we get

$$\begin{aligned}\hat{U}(t) &= 1 + \frac{1}{i\hbar}\int_0^t \hat{H}(t_1)\left(1 + \frac{1}{i\hbar}\int_0^{t_1} \hat{H}(t_2)\hat{U}(t_2)\mathrm{d}t_2\right)\mathrm{d}t_1 \\ &= 1 + \frac{1}{i\hbar}\int_0^t \hat{H}(t_1)\mathrm{d}t_1 + \left(\frac{1}{i\hbar}\right)^2\int_0^t \hat{H}(t_1)\left(\int_0^{t_1} \hat{H}(t_2)\hat{U}(t_2)\mathrm{d}t_2\right)\mathrm{d}t_1 \\ &= 1 + \frac{1}{i\hbar}\int_0^t \mathrm{d}t_1\hat{H}(t_1) + \left(\frac{1}{i\hbar}\right)^2\int_0^t \mathrm{d}t_2\int_0^{t_2} \mathrm{d}t_1\hat{H}(t_2)\hat{H}(t_1)\hat{U}(t_1), \quad (1.24)\end{aligned}$$

where t_1 and t_2 replaced each other in the third term.

By repeating again this procedure, we get

$$\begin{aligned}\hat{U}(t) = 1 &+ \frac{1}{i\hbar}\int_0^t \mathrm{d}t_1\hat{H}(t_1) + \left(\frac{1}{i\hbar}\right)^2\int_0^t \mathrm{d}t_2\int_0^{t_2} \mathrm{d}t_1\hat{H}(t_2)\hat{H}(t_1) \\ &+ \left(\frac{1}{i\hbar}\right)^3\int_0^t \mathrm{d}t_3\int_0^{t_3} \mathrm{d}t_2\int_0^{t_2} \mathrm{d}t_1\hat{H}(t_3)\hat{H}(t_2)\hat{H}(t_1)\hat{U}(t_1). \quad (1.25)\end{aligned}$$

By repeating this procedure indefinitely, we finally get

$$\begin{aligned}\hat{U}(t) &= \sum_{n=0}^{\infty}\left(\frac{1}{i\hbar}\right)^n\int_0^t \mathrm{d}t_n\int_0^{t_n} \mathrm{d}t_{n-1}\cdots\int_0^{t_2} \mathrm{d}t_1\hat{H}(t_n)\hat{H}(t_{n-1})\cdots\hat{H}(t_1) \\ &= \exp_+\left(\frac{1}{i\hbar}\int_0^t \hat{H}(t')\mathrm{d}t'\right), \quad (1.26)\end{aligned}$$

where $\exp_+(x)$ is called the time-ordered exponential.

Consider a small time period $\Delta t = t/N$, the time-ordered integral is expressed by

$$\exp_+\left(\frac{1}{i\hbar}\int_0^t \hat{H}(t')\mathrm{d}t'\right) = \lim_{N\to\infty}\prod_{n=0}^{N-1}\exp_+\left(\frac{1}{i\hbar}\hat{H}(n\Delta t)\right). \quad (1.27)$$

This can be understood as the Hamiltonian is constant during the small period Δt.

1.4.2 Heisenberg Picture

The linear operator of a physical quantity is time dependent in the Heisenberg picture, while in the Schrödinger picture, the operator is time independent and the state vector is time dependent. The expectation value of physical observable $\hat{A}$ at time t is obtained in the Schrödinger representation

$$\langle A\rangle_t = \langle\varphi(t)|\hat{A}|\varphi(t)\rangle = \langle\varphi(0)|\hat{U}^\dagger(t)\hat{A}\hat{U}(t)|\varphi(0)\rangle. \tag{1.28}$$

If we set the time-dependent operator $\hat{A}_{\mathrm{H}}(t)$ and the time-independent state vector $|\varphi\rangle_{\mathrm{H}}$ as

$$\hat{A}_{\mathrm{H}}(t) = \hat{U}^\dagger(t)\hat{A}\hat{U}(t) \tag{1.29}$$

$$|\varphi\rangle_{\mathrm{H}} = |\varphi(0)\rangle, \tag{1.30}$$

where the subscript H denotes the Heisenberg representation, we get

$$\langle A\rangle_t = \langle\varphi_{\mathrm{H}}|\hat{A}_{\mathrm{H}}|\varphi_{\mathrm{H}}\rangle. \tag{1.31}$$

Therefore, the expectation value at time t can be expressed by using the time-dependent linear operator and the time-independent state vector. In the Heisenberg picture, the time evolution of $\hat{A}_{\mathrm{H}}(t)$ is expressed by[4]

$$i\hbar\frac{\mathrm{d}}{\mathrm{d}t}\hat{A}_{\mathrm{H}}(t) = \left[\hat{A}_{\mathrm{H}}(t), \hat{H}\right]. \tag{1.32}$$

1.4.3 Interaction Picture

The interaction picture has characteristics somewhat intermediate representation between the Schrödinger and Heisenberg representation. In this representation, we separate the Hamiltonian into two parts $\hat{H} = \hat{H}_0 + \hat{H}_{\mathrm{I}}$, where $\hat{H}_0$ is the main Hamiltonian for the system and $\hat{H}_{\mathrm{I}}$ is for an interaction between the system and its environment. In addition, $\hat{H}_0$ is time independent and $\hat{H}_{\mathrm{I}}$ is time dependent.[5] This is the case for a system that interacts with light, which is an oscillating electromagnetic field. A physical observable operator in the interaction picture $\hat{A}_{\mathrm{I}}(t)$ and a state vector $|\varphi_{\mathrm{I}}(t)\rangle$ are defined by

[4]The commutator between $\hat{A}$ and $\hat{B}$ is defined as $\left[\hat{A}, \hat{B}\right] = \hat{A}\hat{B} - \hat{B}\hat{A}$.

[5]The interaction picture is useful to investigate light–matter interaction and nonlinear spectroscopy, because the system interacts with an oscillating electromagnetic filed of light.

$$\hat{A}_{\mathrm{I}}(t) = \exp\left(\frac{i}{\hbar}\hat{H}_0 t\right)\hat{A}\exp\left(\frac{-i}{\hbar}\hat{H}_0 t\right) \tag{1.33}$$

$$|\varphi_{\mathrm{I}}(t)\rangle = \exp\left(\frac{i}{\hbar}\hat{H}_0 t\right)|\varphi(t)\rangle. \tag{1.34}$$

Using the Schrödinger equation, we get

$$i\hbar\frac{\mathrm{d}}{\mathrm{d}t}|\varphi_{\mathrm{I}}(t)\rangle = \hat{H}_{\mathrm{I}}(t)|\varphi_{\mathrm{I}}(t)\rangle. \tag{1.35}$$

The solution is formally

$$|\varphi_{\mathrm{I}}(t)\rangle = \exp\left(\frac{i}{\hbar}\hat{H}_0 t\right)\exp\left(\frac{-i}{\hbar}\hat{H}t\right)|\varphi(0)\rangle. \tag{1.36}$$

In addition, $\hat{A}_{\mathrm{I}}(t)$ obeys the equation

$$\frac{\mathrm{d}}{\mathrm{d}t}\hat{A}_{\mathrm{I}}(t) = \frac{i}{\hbar}\left[\hat{H}_0, \hat{A}_{\mathrm{I}}(t)\right]. \tag{1.37}$$

The interaction picture can be understood as $\hat{H}_0$ being treated in the Heisenberg picture and the interaction part $\hat{H}_{\mathrm{I}}$ in the Schrödinger picture.

1.4.3.1 Perturbative Expansion

When the Hamiltonian has a time-dependent perturbative interaction $\hat{H}(t) = \hat{H}_0 + \hat{H}'(t)$, time evolution of the ket vector is expressed by (1.34) in the interaction representation. The solution is obtained formally as

$$|\varphi_{\mathrm{I}}(t)\rangle = |\varphi_{\mathrm{I}}(t_0)\rangle + \sum_{n=1}^{\infty}\left(\frac{-i}{\hbar}\right)^n \int_{t_0}^{t}\mathrm{d}t_n \int_{t_0}^{t_n}\mathrm{d}t_{n-1}\cdots\int_{t_0}^{t_2}\mathrm{d}t_1$$
$$\hat{H}'_{\mathrm{I}}(t_n)\hat{H}'_{\mathrm{I}}(t_{n-1})\cdots\hat{H}'_{\mathrm{I}}(t_1)|\varphi_{\mathrm{I}}(t_0)\rangle, \tag{1.38}$$

where the initial state is $|\varphi_{\mathrm{I}}(t_0)\rangle$. The time-evolution operator from t_0 to t is defined as $\hat{U}(t, t_0) \equiv \exp(-i\hat{H}_0(t - t_0)/\hbar)$. The ket vector in the Schrödinger representation is expressed as $|\varphi(t)\rangle = \hat{U}(t, t_0)|\varphi_{\mathrm{I}}(t)\rangle$ and $|\varphi(t_0)\rangle = |\varphi_{\mathrm{I}}(t_0)\rangle$. Using these relations, we get

$$\hat{U}^{\dagger}(t, t_0)|\varphi(t)\rangle = |\varphi(t_0)\rangle + \sum_{n=1}^{\infty}\left(\frac{-i}{\hbar}\right)^n \int_{t_0}^{t}\mathrm{d}t_n \int_{t_0}^{t_n}\mathrm{d}t_{n-1}\cdots\int_{t_0}^{t_2}\mathrm{d}t_1$$
$$\hat{H}'_{\mathrm{I}}(t_n)\hat{H}'_{\mathrm{I}}(t_{n-1})\cdots\hat{H}'_{\mathrm{I}}(t_1)|\varphi_{\mathrm{I}}(t_0)\rangle, \tag{1.39}$$

and then

$$
\begin{aligned}
|\varphi(t)\rangle &= \hat{U}(t,t_0)|\varphi(t_0)\rangle + \sum_{n=1}^{\infty}\left(\frac{-i}{\hbar}\right)^n \int_{t_0}^{t} \mathrm{d}t_n \int_{t_0}^{t_n} \mathrm{d}t_{n-1}\cdots\int_{t_0}^{t_2}\mathrm{d}t_1 \\
&\quad \hat{U}(t,t_0)\hat{H}_{\mathrm{I}}'(t_n)\hat{H}_{\mathrm{I}}'(t_{n-1})\cdots\hat{H}_{\mathrm{I}}'(t_1)|\varphi_{\mathrm{I}}(t_0)\rangle \\
&= \hat{U}(t,t_0)|\varphi(t_0)\rangle + \sum_{n=1}^{\infty}\left(\frac{-i}{\hbar}\right)^n \int_{t_0}^{t} \mathrm{d}t_n \int_{t_0}^{t_n} \mathrm{d}t_{n-1}\cdots\int_{t_0}^{t_2}\mathrm{d}t_1 \\
&\quad \hat{U}(t,t_0)\hat{U}^{\dagger}(t_n,t_0)\hat{H}'(t_n)\hat{U}(t,t_n)\hat{U}^{\dagger}(t_{n-1},t_0)\hat{H}'(t_{n-1}) \\
&\quad \cdots\hat{H}'(t_2)\hat{U}(t_2,t_0)\hat{U}^{\dagger}(t_1,t_0)\hat{H}'(t_1)\hat{U}(t_1,t_0)|\varphi(t_0)\rangle.
\end{aligned}
\tag{1.40}
$$

By using the relation: $\hat{U}(t_n,t_{n-1}) = \hat{U}(t_n,t_0)\hat{U}(t_0,t_{n-1}) = \hat{U}(t_n,t_0)\hat{U}^{\dagger}(t_{n-1},t_0)$, the above equation can be expressed as

$$
\begin{aligned}
|\varphi(t)\rangle &= \hat{U}(t,t_0)|\varphi(t_0)\rangle + \sum_{n=1}^{\infty}\left(\frac{-i}{\hbar}\right)^n \int_{t_0}^{t} \mathrm{d}t_n \int_{t_0}^{t_n} \mathrm{d}t_{n-1}\cdots\int_{t_0}^{t_2}\mathrm{d}t_1 \\
&\quad \hat{U}(t,t_n)\hat{H}'(t_n)\hat{U}(t_n,t_{n-1})\hat{H}'(t_{n-1}) \\
&\quad \cdots\hat{H}'(t_2)\hat{U}(t_2,t_1)\hat{H}'(t_1)\hat{U}(t_1,t_0)|\varphi(t_0)\rangle.
\end{aligned}
\tag{1.41}
$$

This indicates that the system propagates with $\hat{U}(t_1,t_0)$ from t_0 under the time-independent Hamiltonian $\hat{H}_0$ until time t_1, when the system interacts with the interaction Hamiltonian (perturbation) $\hat{H}'(t_1)$. After the interaction, the system propagates with $\hat{U}(t_2,t_1)$ until time t_2 and this process repeats until time t. $\hat{U}(t,t_0)|\varphi(t_0)\rangle$ is the zeroth-order perturbation ket vector, and each term in the summation is the nth-order one. For example, the second-order perturbation is

$$
\begin{aligned}
|\varphi^{(2)}(t)\rangle &= \left(\frac{-i}{\hbar}\right)^2 \int_{t_0}^{t}\mathrm{d}t_2\int_{t_0}^{t_2}\mathrm{d}t_1 \\
&\quad \hat{U}(t,t_2)\hat{H}'(t_2)\hat{U}(t_2,t_1)\hat{H}'(t_1)\hat{U}(t_1,t_0)|\varphi(t_0)\rangle.
\end{aligned}
\tag{1.42}
$$

This is shown graphically in a single-sided Feynman diagram (Fig. 1.1) [4], where time flows from the left side to the right side.

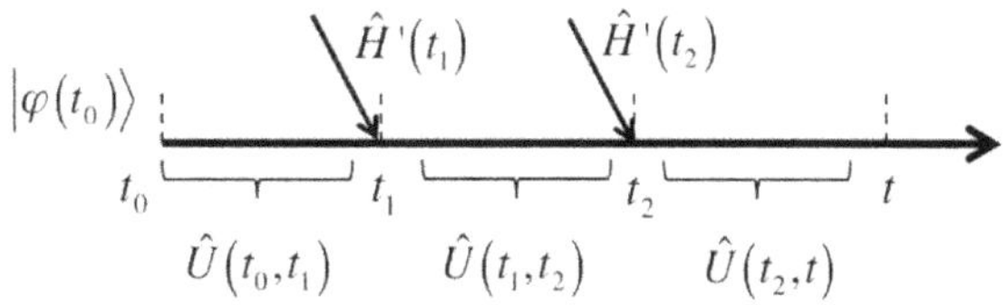

Fig. 1.1 Single-sided Feynman diagram for the second-order perturbation

1.5 Summary

The quantum state is expressed using the ket and bra vectors. Dynamical variables are represented by linear operators that act on ket vectors. Time evolution of the quantum state is expressed by using three ways: the Schrödinger, Heisenberg, and interaction pictures. The time-evolution operator $\hat{U}(t)$, which is a unitary operator, is introduced, and the ket at time t is expressed by $|\Psi(t)\rangle = \hat{U}(t)|\Psi(0)\rangle$. The perturbative expansion is used to express the time evolution of the ket vector in the interaction picture and is shown graphically using the single-sided Feynman diagram [5].

References

1. Sakurai, J.J.: Modern Quantum Mechanics, Revised edn. Addison-Wesley Publishing Company Inc., Reading (1994)
2. Dirac, P.A.M.: The Principles of Quantum Mechanics, 4th edn. Oxford University Press, Oxford (1958)
3. Tannor, D.J.: Introduction to Quantum Mechanics, A Time-Dependent Perspective. University Science Books, California (2007)
4. Hamm, P.: Principles of Nonlinear Optical Spectroscopy: A Practical Approach. http://www.mitr.p.lodz.pl/evu/lectures/Hamm.pdf
5. I wrote this chapter referencing Refs. [1–4] and following books: Takahiro Sunagawa and Masahiro Ueda, *Ryoshi Sokutei to Ryoshi Seigyo (in Japanese, Quantum Measurement and Quantum Control)*, Science Sya (2016); Yoshio Kuramoto and Junichi Ezawa, *Ryoushi Rikigaku (in Japanese, Quantum Mechanics)*, Asakura Shyoten (2008); Keiji Igi and Hikari Kawai, *Kiso Ryousi Rikigaku (in Japanese, Fundamental Quantum Mechanics)*, Kodansya (2007); Masahito Ueda, *Gendai Ryoushi Buturigaku (in Japanese, Modern Quantum Physics)*, Baifukan (2004); Akira Shimizu, *Shin-han Rryoushi Ron no Kiso (in Japanese, New edition Fundamental of Quantum Physics)*, Science Shya (2004)

Chapter 2
Density Operator

This chapter describes the density operator, which is used to calculate light-matter interaction and the generation of coherent phonons in Chap. 6. The density operator can represent both a pure quantum state and a mixed state. The von Neumann equation for the time evolution of the density operator is derived. The perturbative expansion of its time evolution is explained and represented with double-sided Feynman diagrams. Time evolution in a two-level system is shown as an example.

2.1 Pure State and Mixed State

A pure state is a state in which a system is defined by a state vector $|\varphi\rangle$ with a unit probability. On the other hand, when the system includes several quantum states $|\varphi_i\rangle$ with a ratio of p_i, the state is called a mixed state. This state is quite different from a superposition state ($\sum_i \sqrt{p_i}|\varphi_i\rangle$) and cannot be expressed by a ket vector. A mixed state is expressed by a density operator. The mixed state is important when we treat a quantum system consisting of many particles. In this system, some particles are often in different quantum states and the system is expressed with a density operator. On the other hand, when all particles are in a single state, the system is in a pure state.

2.2 Density Operator

2.2.1 *Definition of Density Operator*

When a system consists of state vectors $|\varphi_i\rangle$ with probabilities of p_i, the density operator $\hat{\rho}$ of this system is defined by

K. Nakamura, *Quantum Phononics*, Springer Tracts in Modern Physics 282,
https://doi.org/10.1007/978-3-030-11924-9_2

$$\hat{\rho} \equiv \sum_i p_i |\varphi_i\rangle\langle\varphi_i|, \tag{2.1}$$

where $\sum_i p_i = 1$ [1, 2]. The density operator is expressed by an outer product of ket and bra vectors, which corresponds to a matrix. The probability that we detect the system in the state $|\varphi_i\rangle$ is expressed by $\langle\varphi_i|\hat{\rho}|\varphi_i\rangle$ using the density operator. When a system is in a pure state and has only one state $|\psi\rangle$, the density operator is

$$\hat{\rho} = |\psi\rangle\langle\psi|. \tag{2.2}$$

The density operator is a Hermitian operator because the probability p_i is real and can be expanded with a orthonormal basis set $\{|\varphi_i\rangle\}$. Using an orthonormal basis set, the trace of an arbitrary operator $\hat{A}$ is defined by

$$Tr[\hat{A}] \equiv \sum_i \langle\varphi_i|\hat{A}|\varphi_i\rangle. \tag{2.3}$$

Let us consider a superposition state $|\phi\rangle = \sum_i c_i|\varphi_i\rangle$, where $|\phi\rangle$ is normalized and $\{|\varphi_i\rangle\}$ is a orthonormal basis set. The density operator of this superposition state is calculated as

$$\hat{\rho} = |\phi\rangle\langle\phi| = \sum_{ij} c_i c_j^* |\varphi_i\rangle\langle\varphi_j| = \hat{\rho}' + \sum_{i\neq j} c_i c_j^* |\varphi_i\rangle\langle\varphi_j|, \tag{2.4}$$

where $\hat{\rho}'$ is the density operator of the mixed state composed of $|\varphi_i\rangle$. Then the density operator of the superposition state has additional off-diagonal terms of $c_i c_j^*|\varphi_i\rangle\langle\varphi_j|$ compared to that of the mixed state. These off-diagonal terms are due to interference between the quantum states. A state, which has such a interference term, has "coherence".

Here, we consider, for example, a system including N-particles, in which each particle has only two levels (a ground state $|1\rangle$ and an excited state $|2\rangle$), and N_1 and N_2 particles are in the $|1\rangle$ and $|2\rangle$ states, respectively. Figure 2.1 shows its typical example. When there is no interaction between each particle, the density operator is expressed by

$$\hat{\rho} = \frac{N_1}{N}|1\rangle\langle 1| + \frac{N_2}{N}|2\rangle\langle 2|, \tag{2.5}$$

where $N = N_1 + N_2$. This is an example of the mixed state. On the other hand, in a pure state, all particles are expressed by the single ket vector $|\psi\rangle = \sqrt{N_1/N}|1\rangle + \sqrt{N_2/N}|2\rangle$ and the density operator is expressed by

$$\hat{\rho} = |\psi\rangle\langle\psi| = \frac{N_1}{N}|1\rangle\langle 1| + \frac{N_2}{N}|2\rangle\langle 2| + \sqrt{\frac{N_1 N_2}{N^2}}\left(|1\rangle\langle 2| + |1\rangle\langle 2|\right). \tag{2.6}$$

The last term shows an interference between two states and the coherence.

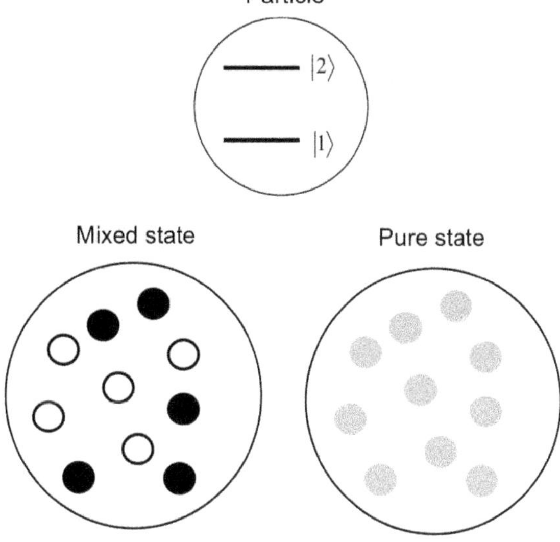

Fig. 2.1 A schematic of pure and mixed states. The particle has only two states ($|1\rangle$ and $|2\rangle$). In the mixed state, particles are in the ground state ($|1\rangle$ represented by black circles) or the excited state ($|2\rangle$ represented by white circles). In the pure state, all particles are in the same state $|\psi\rangle = \sqrt{N_1/N}|1\rangle + \sqrt{N_2/N}|2\rangle$ represented by gray circles

2.2.2 Time Evolution of the Density Operator

We consider a density operator ($\hat{\rho}(t)$) at time t as

$$\hat{\rho}(t) = \sum_i p_i |\varphi_i(t)\rangle\langle\varphi_i(t)|. \tag{2.7}$$

Since $|\varphi_i(t)\rangle$ obeys the Schrödinger equation, the time evolution of the density operator is calculated by

$$\begin{aligned}
\frac{\mathrm{d}}{\mathrm{d}t}\hat{\rho}(t) &= \sum_i p_i \frac{\mathrm{d}}{\mathrm{d}t}|\varphi_i(t)\rangle\langle\varphi_i(t)| \\
&= \sum_i p_i \left(\frac{\mathrm{d}|\varphi_i(t)\rangle}{\mathrm{d}t}\langle\varphi_i(t)| + |\varphi_i(t)\rangle\frac{\mathrm{d}\langle\varphi_i(t)|}{\mathrm{d}t}\right) \\
&= \sum_i p_i \left(-\frac{i}{\hbar}\hat{H}|\varphi_i(t)\rangle\langle\varphi_i(t)| + \frac{i}{\hbar}|\varphi_i(t)\rangle\langle\varphi_i(t)|\hat{H}\right) \\
&= \frac{1}{i\hbar}\left[\hat{H}, \hat{\rho}(t)\right],
\end{aligned} \tag{2.8}$$

which is called the von Neumann equation.[1]

[1]This equation corresponds to the Liouville equation for classical mechanics. Then, this is also called the quantum Liouville equation or Liouville–von Neumann equation [2].

When the Hamiltonian is separated into two parts as $\hat{H} = \hat{H}_0 + \hat{H}'(t)$, where $\hat{H}_0$ is time independent, the density operator can be treated with the interaction picture. The density operator $\hat{\rho}_{\mathrm{I}}(t)$ and the interaction $\hat{H}'_{\mathrm{I}}(t)$ in the interaction picture are given by

$$\hat{\rho}_{\mathrm{I}}(t) = \exp\left(\frac{i\hat{H}_0 t}{\hbar}\right)\hat{\rho}(t)\exp\left(\frac{-i\hat{H}_0 t}{\hbar}\right)$$

$$\hat{H}'_{\mathrm{I}}(t) = \exp\left(\frac{i\hat{H}_0 t}{\hbar}\right)\hat{H}'(t)\exp\left(\frac{-i\hat{H}_0 t}{\hbar}\right), \tag{2.9}$$

and time evolution of the density operator is given by

$$\frac{\mathrm{d}}{\mathrm{d}t}\hat{\rho}_{\mathrm{I}}(t) = \frac{1}{i\hbar}\left[\hat{H}'_{\mathrm{I}}, \hat{\rho}_{\mathrm{I}}(t)\right]. \tag{2.10}$$

The von Neumann equation in the interaction picture is formally the same as that in the Schrödinger picture.

2.2.3 *Perturbative Expansion*

The solution of the von Neumann equation (4.11) is obtained formally as

$$\hat{\rho}(t) = \hat{\rho}(0) + \frac{1}{i\hbar}\int_0^t \left[\hat{H}(t), \hat{\rho}(0)\right]\mathrm{d}t. \tag{2.11}$$

We calculate this integral by using perturbative approximation. At the zeroth order, $\hat{\rho}^{(0)}(t) = \hat{\rho}(0)$. Inserting this into the integration, we get the first-order approximation

$$\hat{\rho}^{(1)}(t) = \frac{1}{i\hbar}\int_0^t \left[\hat{H}(t_1), \hat{\rho}(0)\right]\mathrm{d}t_1. \tag{2.12}$$

This corresponds to that the interaction occurs once at time t_1. Inserting $\hat{\rho}^{(1)}(t)$ into the integral, we get the second-order approximation

$$\hat{\rho}^{(2)}(t) = \left(\frac{1}{i\hbar}\right)^2 \int_0^t \mathrm{d}t_2 \int_0^{t_2} \mathrm{d}t_1 \left[\hat{H}(t_2), \left[\hat{H}(t_1), \hat{\rho}(0)\right]\right]. \tag{2.13}$$

Repeating this procedure one after another, we get the nth-order approximation

$$\hat{\rho}^{(n)}(t) = \left(\frac{1}{i\hbar}\right)^n \int_0^t \mathrm{d}t_n \int_0^{t_n} \mathrm{d}t_{n-1} \int_0^{t_2} \mathrm{d}t_1 \cdots$$
$$\left[\hat{H}(t_n), \left[\hat{H}(t_{n-1}), \ldots \left[\hat{H}(t_1), \hat{\rho}(0)\right] \ldots\right]\right]. \tag{2.14}$$

Finally, we get

$$\hat{\rho}(t) = \sum_{n=0}^{\infty} \hat{\rho}^{(n)}(t). \tag{2.15}$$

The first-order approximation includes a term

$$\begin{aligned}\left[\hat{H}(t_1), \hat{\rho}(0)\right] &= \hat{H}(t_1)\hat{\rho}(0) - \hat{\rho}(0)\hat{H}(t_1) \\ &= \hat{H}(t_1)|\varphi(0)\rangle\langle\varphi(0)| - |\varphi(0)\rangle\langle\varphi(0)|\hat{H}(t_1),\end{aligned} \tag{2.16}$$

which indicates that $|\varphi(0)\rangle$ and $\langle\varphi(0)|$ get interaction once at time t_1. In a similar way, the second-order approximation includes

$$\begin{aligned}\left[\hat{H}(t_2), \left[\hat{H}(t_1), \hat{\rho}(0)\right]\right] &= \hat{H}(t_2)\hat{H}(t_1)|\varphi(0)\rangle\langle\varphi(0)| - \hat{H}(t_2)|\varphi(0)\rangle\langle\varphi(0)|\hat{H}(t_1) \\ &- \hat{H}(t_1)|\varphi(0)\rangle\langle\varphi(0)|\hat{H}(t_2) - |\varphi(0)\rangle\langle\varphi(0)|\hat{H}(t_1)\hat{H}(t_2),\end{aligned} \tag{2.17}$$

which consists of the ket having interaction at t_1 and t_2, the ket having interaction at t_2 with the bra having interaction at t_1, the ket having interaction at t_2 with the bra having interaction at t_1, and the bra having interaction at t_1 and t_2.

2.2.3.1 Perturbative Interaction

The von Neumann equation in the interaction picture is formally equivalent to that in the Schrödinger picture. Then, the density operator in the interaction picture is also expressed as

$$\hat{\rho}_{\mathrm{I}}(t) = \hat{\rho}_{\mathrm{I}}(t_0) + \sum_{n=1}^{\infty} \left(\frac{-i}{\hbar}\right)^n \int_{t_0}^t \mathrm{d}t_n \int_{t_0}^{t_n} \mathrm{d}t_{n-1} \cdots \int_{t_0}^{t_2} \mathrm{d}t_1$$
$$\left[\hat{H}'_{\mathrm{I}}(t_n), \left[\hat{H}'_{\mathrm{I}}(t_{n-1}), \ldots, \left[\hat{H}'_{\mathrm{I}}(t_1), \hat{\rho}_{\mathrm{I}}(t_0)\right] \ldots\right]\right]. \tag{2.18}$$

From this equation, the density operator in the Schrödinger picture is expressed as

$$\hat{\rho}(t) = \hat{U}(t, t_0)\hat{\rho}(t_0)\hat{U}^\dagger(t, t_0) + \sum_{n=1}^{\infty} \left(\frac{-i}{\hbar}\right)^n \int_{t_0}^t \mathrm{d}t_n \int_{t_0}^{t_n} \mathrm{d}t_{n-1} \cdots \int_{t_0}^{t_2} \mathrm{d}t_1$$
$$\hat{U}(t, t_0)\left[\hat{H}'_{\mathrm{I}}(t_n), \left[\hat{H}'_{\mathrm{I}}(t_{n-1}), \ldots, \left[\hat{H}'_{\mathrm{I}}(t_1), \hat{\rho}(t_0)\right] \ldots\right]\right]\hat{U}^\dagger(t, t_0), \tag{2.19}$$

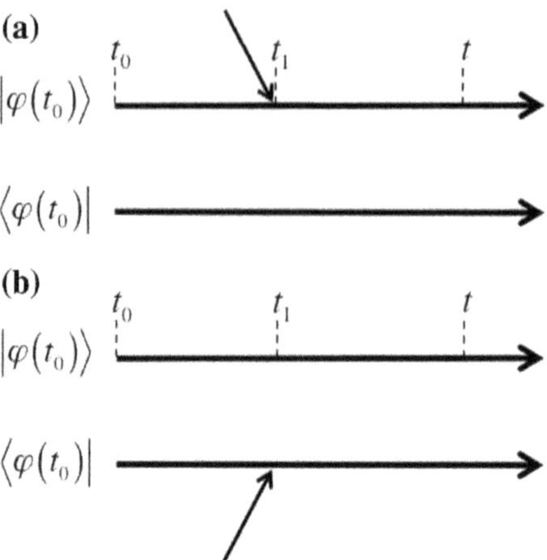

Fig. 2.2 Double-sided Feynman diagrams for the first-order perturbation. **a** for the first term, **b** for the second term in (2.22)

where we used $\hat{\rho}_{\mathrm{I}}(t_0) = \hat{\rho}(t_0)$. The first term is the zeroth-order perturbation and the time-evolution operator $\hat{U}(t, t_0)$ and its Hermitian conjugate operator act on ket and bra vectors, respectively. The first-order perturbation is obtained as

$$\hat{\rho}^{(1)}(t) = \left(\frac{-i}{\hbar}\right) \int_{t_0}^{t} \mathrm{d}t_1 \hat{U}(t, t_0) \left[\hat{H}_{\mathrm{I}}'(t_1), \hat{\rho}(t_0)\right] \hat{U}^{\dagger}(t, t_0). \tag{2.20}$$

The term in the integral is calculated by

$$\begin{aligned} &\hat{U}(t, t_0) \left(\hat{H}_{\mathrm{I}}'(t_1)\hat{\rho}(t_0) - \hat{\rho}(t_0)\hat{H}_{\mathrm{I}}'(t_1)\right) \hat{U}^{\dagger}(t, t_0) \\ &= \hat{U}(t, t_0)\hat{U}^{\dagger}(t_1, t_0)\hat{H}'(t_1)\hat{U}(t_1, t_0)|\varphi(t_0)\rangle\langle\varphi(t_0)|\hat{U}^{\dagger}(t, t_0) \\ &\quad -\hat{U}(t, t_0)|\varphi(t_0)\rangle\langle\varphi(t_0)|\hat{U}^{\dagger}(t_1, t_0)\hat{H}'(t_1)\hat{U}(t_1, t_0)\hat{U}^{\dagger}(t, t_0) \\ &= \hat{U}(t, t_1)\hat{H}'(t_1)\hat{U}(t_1, t_0)|\varphi(t_0)\rangle\langle\varphi(t_0)|\hat{U}^{\dagger}(t, t_0) \\ &\quad -\hat{U}(t, t_0)|\varphi(t_0)\rangle\langle\varphi(t_0)|\hat{U}^{\dagger}(t_1, t_0)\hat{H}'(t_1)\hat{U}^{\dagger}(t, t_1). \end{aligned} \tag{2.21}$$

The first term shows that the ket propagates under $\hat{H}_0$ until time t_1, interacts with perturbation $\hat{H}'(t_1)$ at time t_1, and propagates again under $\hat{H}_0$ and the bra propagates under $\hat{H}_0$ without perturbation. In the second term, the perturbative interaction occurs for the bra. This interpretation is expressed by using double-sided Feynman diagrams (Fig. 2.2). The upper and lower lines show propagation of the ket and bra vectors, respectively. Time is running from the left to the right.

Similar calculation gives the second-order perturbation as

$$\hat{\rho}^{(2)}(t) = \left(\frac{-i}{\hbar}\right)^2 \int_{t_0}^{t} \mathrm{d}t_2 \int_{t_0}^{t_2} \mathrm{d}t_1 \hat{U}(t, t_0) \left[\hat{H}_{\mathrm{I}}'(t_2), \left[\hat{H}_{\mathrm{I}}'(t_1), \hat{\rho}(t_0)\right]\right] \hat{U}^{\dagger}(t, t_0)$$

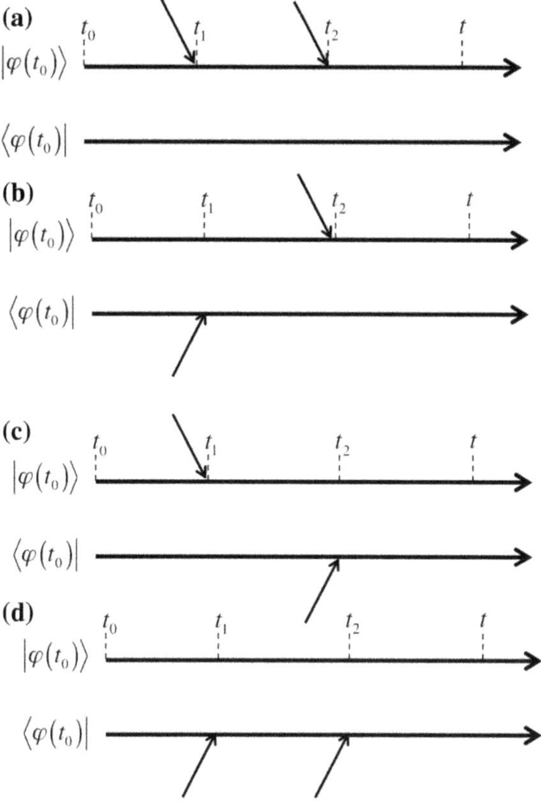

Fig. 2.3 Double-sided Feynman diagrams for the first-order perturbation

$$
\begin{aligned}
&= \left(\frac{-i}{\hbar}\right)^2 \int_{t_0}^{t} \mathrm{d}t_2 \int_{t_0}^{t_2} \mathrm{d}t_1 \\
&\quad (\hat{U}(t,t_2)\hat{H}'(t_2)\hat{U}(t_2,t_1)\hat{H}'(t_1)\hat{U}(t_1,t_0)|\varphi(t_0)\rangle\langle\varphi(t_0)|\hat{U}^\dagger(t,t_0) \\
&\quad - \hat{U}(t,t_2)\hat{H}'(t_2)\hat{U}(t_2,t_0)|\varphi(t_0)\rangle\langle\varphi(t_0)|\hat{U}^\dagger(t_1,t_0)\hat{H}'(t_1)\hat{U}^\dagger(t,t_1) \\
&\quad - \hat{U}(t,t_1)\hat{H}'(t_1)\hat{U}(t_1,t_0)|\varphi(t_0)\rangle\langle\varphi(t_0)|\hat{U}^\dagger(t_2,t_0)\hat{H}'(t_2)\hat{U}^\dagger(t,t_2) \\
&\quad + \hat{U}(t,t_0)|\varphi(t_0)\rangle\langle\varphi(t_0)|\hat{U}^\dagger(t_1,t_0)\hat{H}'(t_1)\hat{U}^\dagger(t_2,t_1)\hat{H}'(t_2)\hat{U}^\dagger(t,t_2)).
\end{aligned} \tag{2.22}
$$

The perturbative interaction occurs two times for the ket or the bra in the first or the fourth term, respectively, and occurs one time for both the ket and bra in the second and third terms. The double-sided Feynman diagrams for the second-order perturbation is shown in Fig. 2.3, in which (a)–(d) correspond each term in the (2.22).

2.2.3.2 Double-Sided Feynman Diagram Rule

The perturbative expansion of the time evolution of the density operator is expressed using double-sided Feynman diagrams. Here, we list several rules for the double-sided Feynman diagram [3].

- Time is running from the left to the right.
- The double-sided Feynman diagram consists of two lines which expressed time evolution of ket and bra vectors.
- Interaction with the interaction Hamiltonian $\hat{H}'(t)$ is represented by arrows.
- Each diagram has a sign $(-1)^n$, where n is the number of interaction to the bra vector (arrows to the bottom line).

2.3 Density Operator for a Two-Level System

As an example of the density operator, here we consider a two-level system which interacts with optical pulses. The system consists of the basis $\{|1\rangle, |2\rangle\}$: $|1\rangle$ and $|2\rangle$ represent the ground and excited states, respectively.

2.3.1 *Interaction-Free Case*

We consider, at first, an interaction-free case. The Hamiltonian is defined by

$$\hat{H} = \varepsilon_1|1\rangle\langle 1| + \varepsilon_2|2\rangle\langle 2|, \tag{2.23}$$

where ε_1 and ε_2 are the energy and $\varepsilon_1 < \varepsilon_2$. The Hamiltonian is also expressed using the matrix:

$$\hat{H} = \begin{pmatrix} \varepsilon_1 & 0 \\ 0 & \varepsilon_2 \end{pmatrix}. \tag{2.24}$$

The time evolution of the density operator, which is shown in the matrix form, is obtained from the von Neumann equation

$$\frac{\mathrm{d}}{\mathrm{d}t}\begin{pmatrix} \rho_{11} & \rho_{12} \\ \rho_{21} & \rho_{22} \end{pmatrix} = \frac{-i}{\hbar}\begin{pmatrix} 0 & (\varepsilon_1 - \varepsilon_2)\rho_{12} \\ (\varepsilon_2 - \varepsilon_1)\rho_{21} & 0 \end{pmatrix}, \tag{2.25}$$

for the interaction-free case. The diagonal components, which correspond to populations, are constant ($\rho_{11}(t) = \rho_{11}(0)$ and $\rho_{22}(t) = \rho_{22}(0)$) in time. The off-diagonal components, which correspond to polarizations, oscillate

$$\rho_{12}(t) = e^{i\omega_{21}t}\rho_{12}(0)$$
$$\rho_{21}(t) = e^{-i\omega_{21}t}\rho_{21}(0), \tag{2.26}$$

where $\omega_{21} = (\varepsilon_2 - \varepsilon_1)/\hbar$.

Using the density operator, phase relaxation can be phenomenologically introduced as

$$\dot{\rho_{12}}(t) = i\omega_{21}\rho_{12}(t) - \Gamma\rho_{12}(t)$$
$$\dot{\rho_{21}}(t) = -i\omega_{21}\rho_{21}(t) - \Gamma\rho_{21}(t), \tag{2.27}$$

where Γ is a relaxation rate, and we get

$$\rho_{12}(t) = e^{i\omega_{21}t}e^{-\Gamma t}\rho_{12}(0)$$
$$\rho_{21}(t) = e^{-i\omega_{21}t}e^{-\Gamma t}\rho_{21}(0). \tag{2.28}$$

2.3.2 *Time-Dependent Interaction Case*

When the oscillating perturbation $\hat{V}(t)$[2]

$$\hat{V}(t) = \gamma e^{-i\omega t}|2\rangle\langle 1| + \gamma e^{i\omega t}|1\rangle\langle 2| \tag{2.29}$$

is applied to the two-level system, the total Hamiltonian ($\hat{H}(t) = \hat{H}_0 + \hat{V}(t)$) is expressed by

$$\hat{H}(t) = \begin{pmatrix} \varepsilon_1 & \gamma e^{i\omega t} \\ \gamma e^{-i\omega t} & \varepsilon_2 \end{pmatrix}. \tag{2.30}$$

The von Neumann equation is

$$\frac{\mathrm{d}}{\mathrm{d}t}\begin{pmatrix} \rho_{11} & \rho_{12} \\ \rho_{21} & \rho_{22} \end{pmatrix} = \frac{-i}{\hbar}\begin{pmatrix} \gamma e^{i\omega t}\rho_{21} - \gamma e^{-i\omega t}\rho_{12} & (\varepsilon_1 - \varepsilon_2)\rho_{12} + \gamma e^{i\omega t}(\rho_{22} - \rho_{11}) \\ (\varepsilon_2 - \varepsilon_1)\rho_{21} + \gamma e^{-i\omega t}(\rho_{11} - \rho_{22}) & \gamma e^{-i\omega t}\rho_{12} - \gamma e^{i\omega t}\rho_{21} \end{pmatrix}. \tag{2.31}$$

[2]This interaction corresponds to an dipole interaction between the two-level system and light with an angular frequency of ω. When we use μ as a dipole moment and E_0 as an amplitude of electric field, $\gamma = \mu E_0$. $\gamma e^{-i\omega t}|2\rangle\langle 1|$ and $\gamma e^{i\omega t}|1\rangle\langle 2|$ correspond to the excitation process with light absorption and the de-excitation process with emission. In addition, the rotating wave approximation is assumed.

By using the frequency $\Omega \equiv \gamma/\hbar$, we get

$$\begin{aligned}
\dot{\rho}_{11} &= -i\Omega(e^{i\omega t}\rho_{21} - e^{-i\omega t}\rho_{12}) \\
\dot{\rho}_{12} &= i\omega_{21}\rho_{12} - i\Omega e^{i\omega t}(\rho_{22} - \rho_{11}) \\
\dot{\rho}_{21} &= -i\omega_{21}\rho_{12} - i\Omega e^{-i\omega t}(\rho_{11} - \rho_{22}) \\
\dot{\rho}_{22} &= -i\Omega(e^{-i\omega t}\rho_{12} - e^{i\omega t}\rho_{21}).
\end{aligned} \tag{2.32}$$

The sum of the population in $|1\rangle$ and $|2\rangle$ is time independent, because $\dot{\rho_{11}} + \dot{\rho_{22}} = 0$. From the results of the interaction-free case, the off-diagonal terms have $e^{-i\omega_{21}}t$ or $e^{i\omega_{21}}t$. Then, we use a rotating frame defined as $\tilde{\rho}_{12} = e^{-i\omega_{21}t}\rho_{12}$, $\tilde{\rho}_{21} = e^{i\omega_{21}t}\rho_{21}$, $\tilde{\rho}_{11} = \rho_{11}$, and $\tilde{\rho}_{22} = \rho_{22}$. In the rotating frame, the density matrix elements are

$$\begin{aligned}
\dot{\tilde{\rho}}_{12} &= -i\Omega e^{i(\omega-\omega_{21})t}(\tilde{\rho}_{22} - \tilde{\rho}_{11}) \\
\dot{\tilde{\rho}}_{21} &= i\Omega e^{-i(\omega-\omega_{21})t}(\tilde{\rho}_{22} - \tilde{\rho}_{11}) \\
\dot{\tilde{\rho}}_{11} &= -i\Omega e^{i(\omega-\omega_{21})t}\tilde{\rho}_{21} + i\Omega e^{-i(\omega-\omega_{21})t}\tilde{\rho}_{12} \\
\dot{\tilde{\rho}}_{22} &= i\Omega e^{i(\omega-\omega_{21})t}\tilde{\rho}_{21} - i\Omega e^{-i(\omega-\omega_{21})t}\tilde{\rho}_{12}.
\end{aligned} \tag{2.33}$$

For the resonant condition ($\omega - \omega_{21} = 0$), the equations are greatly simplified, and we get

$$\begin{aligned}
\dot{\tilde{\rho}}_{12} - \dot{\tilde{\rho}}_{21} &= 2i\Omega(\tilde{\rho}_{11} - \tilde{\rho}_{22}) \\
\dot{\tilde{\rho}}_{11} - \dot{\tilde{\rho}}_{22} &= 2i\Omega(\tilde{\rho}_{12} - \tilde{\rho}_{21}),
\end{aligned} \tag{2.34}$$

and

$$\ddot{\tilde{\rho}}_{11} - \ddot{\tilde{\rho}}_{22} = 2i\Omega(\dot{\tilde{\rho}}_{12} - \dot{\tilde{\rho}}_{21}) = -4\Omega^2(\tilde{\rho}_{11} - \tilde{\rho}_{22}). \tag{2.35}$$

If we put $\tilde{\rho}_{11} - \tilde{\rho}_{22} = Ce^{at}$, we get $a = \pm i2\Omega$. For the initial condition of $\rho_{11}(0) = 1$ and $\rho_{22}(0) = 0$, the coefficient C is obtained to be $1/2$. Then, we get

$$\rho_{11} - \rho_{22} = \frac{e^{i2\Omega t} + e^{-i2\Omega t}}{2} = \cos 2\Omega t. \tag{2.36}$$

By using $\rho_{11} + \rho_{22} = 1$, we finally get

$$\rho_{11} = \frac{1 - \cos 2\Omega t}{2} = \cos^2 \Omega t. \tag{2.37}$$

The result shows that the system absorbs energy from the interaction potential $\hat{V}(t)$ in the time range between 0 and $\Omega/2$ and emits in the time range between $\Omega/2$ and Ω (as shown in Fig. 2.4). This absorption and emission process periodically repeat with a frequency of Ω, which is called Rabi frequency.

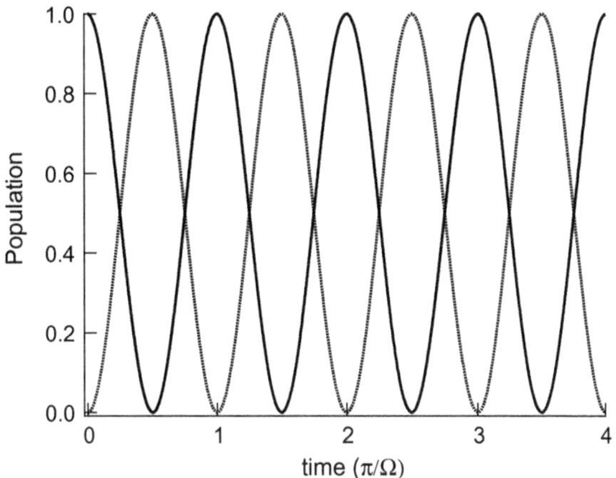

Fig. 2.4 Time evolution of the population (ρ_{11}, and ρ_{22}) of the ground and excited states. The solid and dotted curves show ρ_{11} and ρ_{22}, respectively

When we introduce the phase relaxation (Γ) in this two-level system, $\dot{\tilde{\rho_{12}}}$ and $\dot{\tilde{\rho_{21}}}$ are obtained by

$$\begin{aligned}\dot{\tilde{\rho}}_{12} &= -\Gamma\tilde{\rho_{12}} - i\Omega e^{i(\omega-\omega_{21})t}(\tilde{\rho}_{22} - \tilde{\rho}_{11})\\ \dot{\tilde{\rho}}_{21} &= -\Gamma\tilde{\rho_{21}} i\Omega e^{-i(\omega-\omega_{21})t}(\tilde{\rho}_{22} - \tilde{\rho}_{11}).\end{aligned} \tag{2.38}$$

At resonance, a similar calculation gives

$$\rho_{11} = \frac{1}{2}\left(1 + \frac{e^{-\Gamma t}}{2}\left(e^{\sqrt{\Gamma^2-4\Omega^2}t} + e^{-\sqrt{\Gamma^2-4\Omega^2}t}\right)\right). \tag{2.39}$$

If we represent the relaxation rate in units of the Rabi oscillation ($\Gamma = n\Omega$), the population ρ_{11} is a damped oscillation

$$\rho_{11} = \frac{1}{2}\left(1 + e^{-n\Omega t}\cos(\sqrt{4-n^2}\Omega t)\right), \tag{2.40}$$

if $0 \le n \le 2$ and

$$\rho_{11} = \frac{1}{2}\left(1 + e^{-n\Omega t}\cosh(\sqrt{n^2-4}\Omega t)\right), \tag{2.41}$$

if $n > 2$.

The time evolution of the population (ρ_{11} and ρ_{22}) is shown in Figs. 2.5 and 2.6 for $\Gamma = 0.5\,\Omega$ and $\Gamma = 2.5\,\Omega$, respectively. Both ρ_{11} and ρ_{22} approach to the steady-state value (1/2).

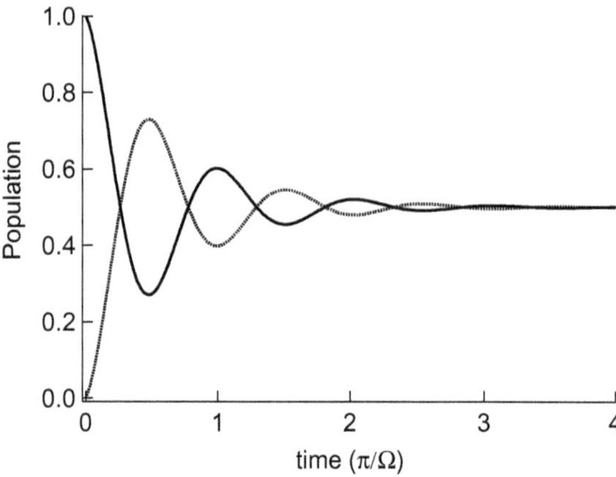

Fig. 2.5 Time evolution of the population (ρ_{11}, and ρ_{22}) of the ground and excited states with relaxation ($\Gamma = 0.5\,\Omega$). The solid and dotted curves show ρ_{11} and ρ_{22}, respectively

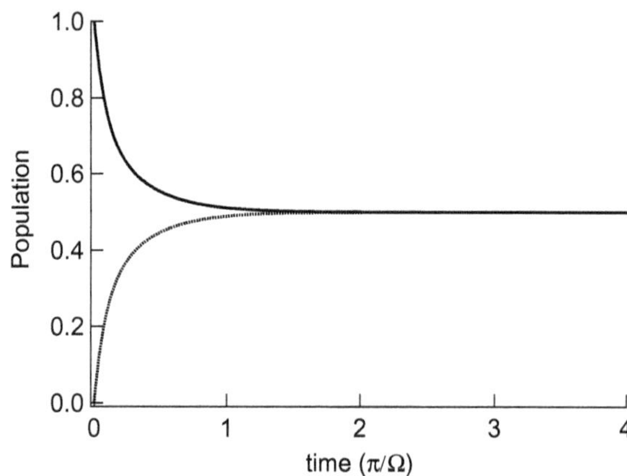

Fig. 2.6 Time evolution of the population (ρ_{11}, and ρ_{22}) of the ground and excited states with relaxation ($\Gamma = 2.5\,\Omega$). The solid and dotted curves show ρ_{11} and ρ_{22}, respectively

2.4 Summary

The density operator, which can describe a pure state and mixed state, is described. The time evolution of the density matrix is obtained by the von Neumann equation. The perturbative expansion of the time evolution of the density operator is explained and represented by using a double-sided Feynman diagram. The time evolution of the populations in the excited and ground states in the two-level model is calculated [4].

References

1. Schatz, G.C., Ratner, M.A.: Quantum Mechanics in Chemistry. Dover Publication Inc., Mineola (2002)
2. Tannor, D.J.: Introduction to Quantum Mechanics, A Time-Dependent Perspective. University Science Books, California (2007)

3. Mukamel, S.: Principles of Nonlinear Optical Spectroscopy. Oxford University Press, New York (1995)
4. I wrote this chapter referencing Refs. [1–3] and following books: Yoshio Kuramoto and Junichi Ezawa, *Ryoushi Rikigaku (in Japanese, Quantum Mechanics)*, Asakura Shyoten (2008); Keiji Igi and Hikari Kawai, *Kiso Ryousi Rikigaku (in Japanese, Fundamental Quantum Mechanics)*, Kodansya (2007); Takahiro Sunagawa and Masahiro Ueda, *Ryoshi Sokutei to Ryoshi Seigyo (in Japanese, Quantum Measurement and Quantum Control)*, Science Sya (2016), Masahito Ueda *Gendai Ryoushi Buturigaku (in Japanese, Modern Quantum Physics)*, Baifukan (2004); Akira Shimizu, *Shin-han Rryoushi Ron no Kiso (in Japanese, New edition Fundamental of Quantum Physics)*, Science Shya (2004); Kyo Inoue, *Kogaku Kei no Tameno Ryoushi Kogaku (in Japanese, Quantum Optics for Engineer)*, Morikita Shyoten (2015); Masahiro Matsuoka, *Ryoushi Kogaku (in Japanese, Quantum Optics)*, Shokabou (2000)

Chapter 3
Harmonic Oscillator and Coherent and Squeezed States

This chapter describes the quantum mechanics of a harmonic oscillator, which is of essential importance in treating a phonon, using creation and annihilation operators. A number state, a coherent state, and a squeezed state are introduced. The number state is an eigenstate of the energy of the harmonic oscillator. The coherent state is a minimum uncertainty state. In the squeezed state, fluctuation of one of the conjugated variables is reduced and lowered than that of the vacuum state. Coherent and squeezed states are important in quantum optics and phononics. Time evolution of the mean value of position and variance for the harmonic oscillator is discussed.

3.1 Hamiltonian and Energy Eigenstate

3.1.1 Hamiltonian

The classical Hamiltonian H of a one-dimensional harmonic oscillator with a mass m and an angular frequency ω is

$$H = \frac{p^2}{2m} + \frac{m\omega^2 x^2}{2}, \tag{3.1}$$

where x and p are position and momentum, respectively (Fig. 3.1).

Replacing x and p to linear operators $\hat{x}$ and $\hat{p}$, which satisfy commutation relation $\left[\hat{x}, \hat{p}\right] = i\hbar$, we get the quantum mechanical Hamiltonian $\hat{H}$ by

$$\hat{H} = \frac{\hat{p}^2}{2m} + \frac{m\omega^2 \hat{x}^2}{2}. \tag{3.2}$$

K. Nakamura, *Quantum Phononics*, Springer Tracts in Modern Physics 282,
https://doi.org/10.1007/978-3-030-11924-9_3

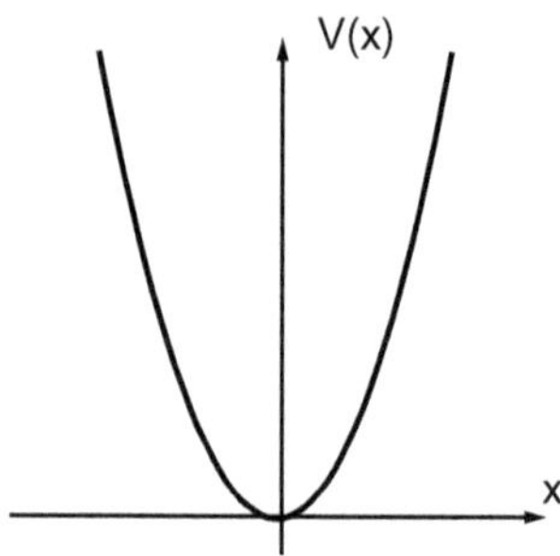

Fig. 3.1 Potential energy of the harmonic oscillator

Here, we introduce new operators $\hat{a}$ and $\hat{a}^\dagger$ [1], which are defined using $\hat{x}$ and $\hat{p}$ by

$$\begin{aligned}\hat{a} &\equiv \frac{1}{\sqrt{2m\hbar\omega}}\left(m\omega\hat{x} + i\hat{p}\right), \\ \hat{a}^\dagger &\equiv \frac{1}{\sqrt{2m\hbar\omega}}\left(m\omega\hat{x} - i\hat{p}\right). \end{aligned} \tag{3.3}$$

The position and momentum operators ($\hat{x}$ and $\hat{p}$) are expressed by

$$\begin{aligned}\hat{x} &= \sqrt{\frac{\hbar}{2m\omega}}\left(\hat{a} + \hat{a}^\dagger\right), \\ \hat{p} &= i\sqrt{\frac{m\hbar\omega}{2}}\left(-\hat{a} + \hat{a}^\dagger\right), \end{aligned} \tag{3.4}$$

using the new operators. Inserting this relationship into the commutation relation $\left[\hat{x}, \hat{p}\right]$, we get

$$\begin{aligned}\left[\hat{x}, \hat{p}\right] &= \hat{x}\hat{p} - \hat{p}\hat{x} \\ &= \frac{-i\hbar}{2}\left((\hat{a} + \hat{a}^\dagger)(\hat{a} - \hat{a}^\dagger) - (\hat{a} - \hat{a}^\dagger)(\hat{a} + \hat{a}^\dagger)\right) \\ &= -i\hbar\left(\hat{a}^\dagger\hat{a} - \hat{a}\hat{a}^\dagger\right) \\ &= i\hbar\left[\hat{a}, \hat{a}^\dagger\right]. \end{aligned} \tag{3.5}$$

Then, the commutation relation between $\hat{a}$ and $\hat{a}^\dagger$ is

$$\left[\hat{a}, \hat{a}^\dagger\right] = 1, \tag{3.6}$$

because $\left[\hat{x}, \hat{p}\right] = i\hbar$. The operator $\hat{a}$ is not a Hermitian operator. The Hamiltonian is expressed by using $\hat{a}$ and $\hat{a}^\dagger$ by

$$
\begin{aligned}
\hat{H} &= \frac{1}{2m}\left(i\sqrt{\frac{m\hbar\omega}{2}}\left(-\hat{a}+\hat{a}^{\dagger}\right)\right)^2 + \frac{m\omega^2}{2}\left(\sqrt{\frac{\hbar}{2m\omega}}\left(\hat{a}+\hat{a}^{\dagger}\right)\right)^2 \\
&= \frac{\hbar\omega}{2}\left(\hat{a}\hat{a}^{\dagger}+\hat{a}^{\dagger}\hat{a}\right) \\
&= \hbar\omega\left(\hat{a}^{\dagger}\hat{a}+\frac{1}{2}\right).
\end{aligned}
\tag{3.7}
$$

If we introduce a new Hermitian operator $\hat{n}$[1] by

$$
\hat{n} \equiv \hat{a}^{\dagger}\hat{a}, \tag{3.8}
$$

the Hamiltonian is represented by a very simple form

$$
\hat{H} = \hbar\omega\left(\hat{n}+\frac{1}{2}\right), \tag{3.9}
$$

using a single linear operator $\hat{n}$.

3.1.1.1 Eigenenergy and Energy Eigenstates

Because the Hamiltonian $\hat{H}$ of the harmonic oscillator is represented by the single linear operator $\hat{n}$, it is enough to solve an eigenvalue equation for $\hat{n}$ to obtain an energy eigenvalue and eigenstate. Let us consider that the eigenvalue equation holds for $\hat{n}$ as

$$
\hat{n}|n\rangle = n|n\rangle, \tag{3.10}
$$

where n and $|n\rangle$ are eigenvalue and eigenstate, respectively. Operating $\hat{a}$ on (3.10) from the left side, the left-hand side of the equation is

$$
\hat{a}\hat{n}|n\rangle = \hat{a}\hat{a}^{\dagger}\hat{a}|n\rangle = \left(\hat{a}^{\dagger}\hat{a}+1\right)|n\rangle = \left(\hat{n}+1\right)|n\rangle, \tag{3.11}
$$

and the right-hand side of the equation is

$$
\hat{a}n|n\rangle = n\hat{a}|n\rangle. \tag{3.12}
$$

Then, we get

$$
\left(\hat{n}+1\right)|n\rangle = n\hat{a}|n\rangle, \tag{3.13}
$$

[1] $\hat{n}^{\dagger} = (\hat{a}^{\dagger}\hat{a})^{\dagger} = (\hat{a})^{\dagger}(\hat{a}^{\dagger})^{\dagger} = \hat{a}^{\dagger}\hat{a} = \hat{n}$, because $(\hat{A}\hat{B})^{\dagger} = \hat{B}^{\dagger}\hat{A}^{\dagger}$.

and

$$\hat{n}\hat{a}|n\rangle = (n-1)\hat{a}|n\rangle. \tag{3.14}$$

This equation means that $\hat{a}|n\rangle$ is also an eigenstate of the operator $\hat{n}$ and its eigenvalue is $n-1$. We named an eigenstate belonging to the eigenvalue n $|n\rangle$, then an eigenstate belonging to the eigenvalue $n-1$ is $|n-1\rangle$. The state $\hat{a}|n\rangle$ is expressed by

$$\hat{a}|n\rangle = C_n|n-1\rangle, \tag{3.15}$$

where C_n is a constant. When we calculate the inner product with itself, the left-hand side is

$$\langle n|\hat{a}^{\dagger}\hat{a}|n\rangle = \langle n|\hat{n}|n\rangle = n, \tag{3.16}$$

and the right-hand side is

$$\langle n-1|C_n^* C_n|n-1\rangle = |C_n|^2 \geq 0. \tag{3.17}$$

Then, we find that $n \geq 0$ and $C_n = \pm\sqrt{n}$. We choose a positive number, because the global phase can be set arbitrarily. Equation (3.15) becomes

$$\hat{a}|n\rangle = \sqrt{n}|n-1\rangle. \tag{3.18}$$

n should be a nonnegative integer: $n = 0, 1, 2, 3, \ldots$. We explain this below [2]. The operator $\hat{a}$ changes an eigenstate to another eigenstate belonging to the eigenvalue reduced by one. Operating this operator m times to the eigenstate $|n\rangle$, we get the eigenstate $|n-m\rangle$ as

$$(\hat{a})^m |n\rangle = \sqrt{n(n-1)\cdots(n-m+1)}|n-m\rangle. \tag{3.19}$$

The eigenvalue is nonnegative and $n \geq m$. Suppose an integer m_0 which satisfies $n \geq m_0 > n-1$, the eigenstates should be

$$(\hat{a})^{m_0)} |n\rangle \neq 0, \tag{3.20}$$

and

$$(\hat{a})^{m_0+1)} |n\rangle = 0. \tag{3.21}$$

If (3.21) does not hold, we get a negative eigenvalue $(n-(m_0+1))$. By operating $\hat{a}^{\dagger}$ to (3.21) from the left side, we get

$$\hat{a}^{\dagger} (\hat{a})^{m_0+1} |n\rangle = (\hat{a}^{\dagger}\hat{a}) (\hat{a})^{m_0} |n\rangle = \hat{n} (\hat{a})^{m_0} |n\rangle = 0. \tag{3.22}$$

Using (3.19), we get

$$\hat{n}\left(\hat{a}\right)^{m_0}|n\rangle = (n - m_0)\sqrt{n(n-1)\cdots(n-m+1)}|n-m\rangle = 0. \tag{3.23}$$

Then, we get $n = m_0$. Since m_0 is a nonnegative integer, the eigenvalue n is a nonnegative integer.[2] In conclusion, $|n\rangle$ is an eigenstate of both the operator $\hat{n}$ and the Hamiltonian $\hat{H}$, then we get the eigenvalue equation

$$\hat{H}|n\rangle = \hbar\omega\left(\hat{n} + \frac{1}{2}\right)|n\rangle = \hbar\omega\left(n + \frac{1}{2}\right)|n\rangle, \tag{3.24}$$

where n is a nonnegative integer ($n = 0, 1, 2, \ldots$). Eigenenergies E_n for the eigenstate $|n\rangle$ is

$$E_n = \hbar\omega\left(n + \frac{1}{2}\right). \tag{3.25}$$

The harmonic oscillator has energy of $\hbar\omega/2$ even at the lowest state $|0\rangle$, which is called the zero-point energy, and a constant energy spacing of $\hbar\omega$. The eigenstate $|n\rangle$ is called an occupation number state,[3] because well-defined numbers of quanta ($\hbar\omega$) are included in the state. $|0\rangle$ is the lowest energy state and called the ground state or the vacuum state.

3.1.2 Annihilation and Creation Operators

The operator $\hat{a}$ shifts a number state $|n\rangle$ to the lower state $|n-1\rangle$, and is called the annihilation operator [1]. On the other hand, its conjugated operator $\hat{a}^\dagger$ is the creation operator, which shifts the state to the upper state $|n+1\rangle$. Operating $\hat{a}^\dagger$ to (3.10) from the left side, the left-hand side is

$$\hat{a}^\dagger\hat{n}|n\rangle = \hat{a}^\dagger\hat{a}^\dagger\hat{a}|n\rangle = \hat{a}^\dagger\left(\hat{a}\hat{a}^\dagger - 1\right)|n\rangle = \left(\hat{n} - 1\right)\hat{a}^\dagger|n\rangle \tag{3.26}$$

and the right-hand side is

$$\hat{a}^\dagger n|n\rangle = n\hat{a}^\dagger|n\rangle. \tag{3.27}$$

Then, we get

$$\hat{n}\hat{a}^\dagger|n\rangle = (n+1)\hat{a}^\dagger|n\rangle, \tag{3.28}$$

which means that $\hat{a}^\dagger|n\rangle$ is also an eigenvector with the eigenvalue of $n+1$. So we can write $\hat{a}^\dagger|n\rangle$ as

---

[2] m is, of course, a nonnegative integer, because it is number of times for the operation.

[3] The occupation number state is also called the Fock state for photons in quantum optics.

$$\hat{a}^\dagger|n\rangle = C_n|n+1\rangle, \tag{3.29}$$

where C_n is a normalized constant. When we calculate the inner product with itself, the left-hand side is

$$\langle n|\hat{a}\hat{a}^\dagger|n\rangle = \langle n|(\hat{n}+1)|n\rangle = n+1, \tag{3.30}$$

and the right-hand side is

$$\langle n+1|C_n^* C_n|n+1\rangle = |C_n|^2 \geq 0. \tag{3.31}$$

Then, we get

$$\hat{a}^\dagger|n\rangle = \sqrt{n+1}|n+1\rangle. \tag{3.32}$$

The creation or annihilation operator shifts an eigenstate upward or downward, respectively. The lowest eigenstate is $|0\rangle$ and obeys

$$\hat{a}|0\rangle = 0. \tag{3.33}$$

Operating $\hat{a}^\dagger$ n times to the lowest state $|0\rangle$, we get upper number states one by one as [1]

$$\begin{aligned}
|1\rangle &= \hat{a}^\dagger|0\rangle \\
|2\rangle &= \frac{1}{\sqrt{2}}\hat{a}^\dagger|1\rangle = \frac{1}{\sqrt{2\cdot 1}}(\hat{a}^\dagger)^2|0\rangle \\
|3\rangle &= \frac{1}{\sqrt{3}}\hat{a}^\dagger|2\rangle = \frac{1}{\sqrt{3\cdot 2\cdot 1}}(\hat{a}^\dagger)^3|0\rangle \\
&\vdots \\
|n\rangle &= \frac{1}{\sqrt{n}}\hat{a}^\dagger|n-1\rangle = \frac{1}{\sqrt{n!}}(\hat{a}^\dagger)^n|0\rangle.
\end{aligned} \tag{3.34}$$

The annihilation and creation operators obey commutation relations with the number operator as $\left[\hat{a},\hat{n}\right] = \hat{a}$ and $\left[\hat{a}^\dagger,\hat{n}\right] = \hat{a}^\dagger$.

3.1.3 Wave Function in a Position Space

The wave function $\varphi_0(x)$ of the lowest state $|0\rangle$ is obtained in a position space by solving (3.33).[4] In the one-dimensional position space, the position and momentum operators are expressed as

---

[4]The wave function $\varphi_(x)$ is obtained by projection of the ket vector $|\varphi\rangle$ into position space and has the relationship $\varphi_(x) = \langle x|\varphi\rangle$.

$$\hat{x} = x$$
$$\hat{p} = -i\hbar\frac{\mathrm{d}}{\mathrm{d}x}, \tag{3.35}$$

and the annihilation operator is represented by

$$\hat{a} = \frac{1}{\sqrt{2m\hbar\omega}}\left(m\omega x + \hbar\frac{\mathrm{d}}{\mathrm{d}x}\right). \tag{3.36}$$

Then, the differential equation for the wave function $\varphi_0(x)$ is

$$\left(m\omega x + \hbar\frac{\mathrm{d}}{\mathrm{d}x}\right)\varphi_0(x) = 0, \tag{3.37}$$

and the solution is

$$\varphi_0(x) = C\exp\left(\frac{-m\omega x^2}{2\hbar}\right). \tag{3.38}$$

The normalization coefficient C is calculated by

$$\int_{-\infty}^{\infty}|\varphi_0(x)|^2\mathrm{d}x = 1, \tag{3.39}$$

and obtained as

$$C = \sqrt{\frac{m\omega}{\hbar\pi}}. \tag{3.40}$$

Then, the wave function $\varphi_0(x)$ is a Gaussian function centered at $x = 0$. The wave functions for the upper number states are obtained by operating the creation operator

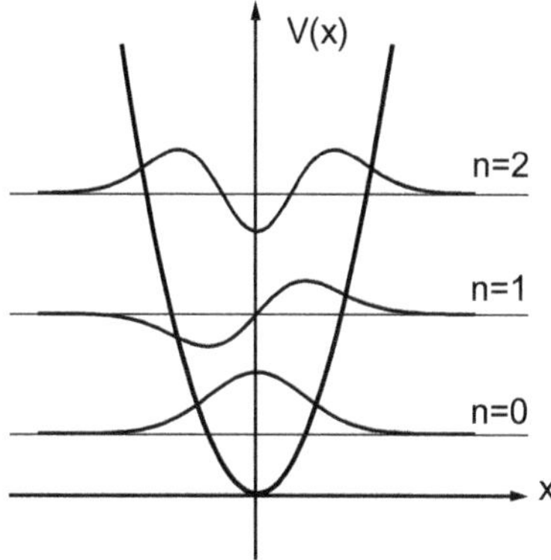

Fig. 3.2 The wave function for $n = 0$, $n = 1$, and $n = 2$ states. E_0, E_1, and E_2 are $\hbar\omega/2$, $3\hbar\omega/2$, and $5\hbar\omega/2$, respectively

$\hat{a}^\dagger$ one by one to $\varphi_0(x)$. For example, the wave function $\varphi_1(x)$ for $|1\rangle$ is

$$\begin{aligned}\varphi_1(x) = \hat{a}^\dagger\varphi_0(x) &= \frac{1}{\sqrt{2m\hbar\omega}}\left(m\omega x - \hbar\frac{\mathrm{d}}{\mathrm{d}x}\right)\left(\sqrt{\frac{m\omega}{\hbar\pi}}\exp\left(\frac{-m\omega x^2}{2\hbar}\right)\right)\\ &= \frac{2m\hbar}{\hbar\sqrt{2\pi}}x\exp\left(\frac{-m\omega x^2}{2\hbar}\right).\end{aligned} \tag{3.41}$$

The wave function $\varphi_1(x)$ is a odd function and has a node at $x = 0$. The $\varphi_2(x)$ is also calculated by $\hat{a}^\dagger\varphi_1(x)$. The wave functions $\varphi_0(x)$, $\varphi_1(x)$, and $\varphi_1(x)$ are shown in Fig. 3.2.

3.2 Time Evolution of a State in the Harmonic Oscillator System

The number state $|n(t)\rangle$ at time t is defined as

$$|n(t)\rangle = \exp\left(-i\left(n+\frac{1}{2}\right)\omega t\right)|n\rangle, \tag{3.42}$$

because its energy is $\hbar\omega(n+1/2)$. Of course, the probability distribution function of the number state does not change in time $\langle n(t)|n(t)\rangle = \langle n|n\rangle$, because it is a stationary state. The expectation value of position for the number state is calculated to be zero:

$$\langle\hat{x}\rangle = \langle n(t)|\hat{x}|n(t)\rangle = \langle n|\left(\sqrt{\frac{\hbar}{2m\omega}}(\hat{a}+\hat{a}^\dagger)\right)|n\rangle = 0. \tag{3.43}$$

The standard deviation σ is obtained by

$$\sigma = \sqrt{\langle\hat{x}^2\rangle - \langle\hat{x}\rangle^2}, \tag{3.44}$$

where $\langle\hat{x}^2\rangle$ is calculated by

$$\begin{aligned}\langle\hat{x}^2\rangle &= \langle n|\left(\sqrt{\frac{\hbar}{2m\omega}}(\hat{a}+\hat{a}^\dagger)\right)^2|n\rangle\\ &= \frac{\hbar}{2m\omega}\langle n|\left(\hat{a}^2+\hat{a}\hat{a}^\dagger+\hat{a}^\dagger\hat{a}+(\hat{a}^\dagger)^2\right)|n\rangle\\ &= \frac{\hbar}{2m\omega}\langle n|\left(2\hat{n}+1\right)|n\rangle = \frac{(2n+1)\hbar}{2m\omega}.\end{aligned} \tag{3.45}$$

Then, we get the standard deviation as

$$\sigma = \sqrt{\frac{(2n+1)\hbar}{2m\omega}}. \tag{3.46}$$

When the position of the harmonic oscillator is measured, the observed value is scattered at every observation and its standard deviation is σ, with an average value of zero. This feature is different from the characteristics of classical harmonic oscillators, in which the position changes in time with its angular frequency ω. Such oscillation of position can be realized in a superposition state of adjoining number states. Suppose the superposition state of $|0\rangle$ and $|1\rangle$ as

$$|\varphi(t)\rangle = \frac{1}{\sqrt{2}}\left(|0\rangle + e^{-i\omega t}|1\rangle\right)e^{-i\frac{\omega}{2}t}. \tag{3.47}$$

The expectation value of position in this state is

$$\begin{aligned}\langle x\rangle &= \frac{1}{2}\left(\langle 0| + e^{i\omega t}\langle 1|\right)\sqrt{\frac{\hbar}{2m\omega}}\left(\hat{a}+\hat{a}^{\dagger}\right)\left(|0\rangle + e^{-i\omega t}|1\rangle\right)\\ &= \sqrt{\frac{\hbar}{2m\omega}}\cos(\omega t).\end{aligned} \tag{3.48}$$

The expectation value of $\langle x^2\rangle$ is obtained by

$$\begin{aligned}\langle x^2\rangle &= \frac{1}{2}\left(\langle 0| + e^{i\omega t}\langle 1|\right)\frac{\hbar}{2m\omega}\left(2\hat{n}+1\right)\left(|0\rangle + e^{-i\omega t}|1\rangle\right)\\ &= \frac{\hbar}{m\omega},\end{aligned} \tag{3.49}$$

and the standard deviation is

$$\begin{aligned}\sigma &= \sqrt{\frac{\hbar}{m\omega} - \frac{\hbar}{2m\omega}\cos^2(\omega t)}\\ &= \sqrt{\frac{\hbar}{2m\omega}(1+\sin^2(\omega t))}.\end{aligned} \tag{3.50}$$

Then, the standard deviation for position also oscillates, which means the shape of the probability distribution function changes in time.

On the other hand, a superposition state composed of number states, which differ more than two, shows no oscillation in expectation value of position, because the position operator includes only $\hat{a}$ and $\hat{a}^{\dagger}$. Suppose the superposition state of $|0\rangle$ and $|2\rangle$ are

$$|\varphi(t)\rangle = \frac{1}{\sqrt{2}}\left(|0\rangle + e^{-i2\omega t}|2\rangle\right)e^{-i\frac{\omega}{2}t}. \tag{3.51}$$

The expectation value of position in the state is

$$\langle x\rangle = \frac{1}{2}\left(\langle 0| + e^{i2\omega t}\langle 2|\right)\sqrt{\frac{\hbar}{2m\omega}}\left(\hat{a}+\hat{a}^\dagger\right)\left(|0\rangle + e^{-2\omega t}|2\rangle\right) = 0. \tag{3.52}$$

The expectation value of $\langle x^2\rangle$ is obtained by

$$\begin{aligned}\langle x^2\rangle &= \frac{1}{2}\left(\langle 0| + e^{i2\omega t}\langle 2|\right)\frac{\hbar}{2m\omega}\left(\hat{a}^2 + (\hat{a}^\dagger)^2 + 2\hat{n} + 1\right)\left(|0\rangle + e^{-i2\omega t}|2\rangle\right)\\ &= \frac{\hbar}{2m\omega}\left(3+\sqrt{2}\cos(2\omega t)\right),\end{aligned} \tag{3.53}$$

and the standard deviation is the same value

$$\sigma = \frac{\hbar}{2m\omega}\left(3+\sqrt{2}\cos(2\omega t)\right). \tag{3.54}$$

Therefore, in the superposition state, the mean value does not change and only the standard deviation oscillates with 2ω.

3.3 Coherent States

3.3.1 Definition of the Coherent State

A coherent state[5] is one of the most important states, which is used in quantum optics and quantum information technologies [4]. The coherent state is defined by an eigenstate of the annihilation operator $\hat{a}$. Consider an eigenvalue and eigenstate of $\hat{a}$ as α and $|\alpha\rangle$

$$\hat{a}|\alpha\rangle = \alpha|\alpha\rangle. \tag{3.55}$$

The eigenvalue α is, in general, complex, because $\hat{a}$ is not Hermitian. When we expand this coherent state using number states $|n\rangle$, we get

$$|\alpha\rangle = \sum_n C_n|n\rangle, \tag{3.56}$$

where $C_n = \langle n|\alpha\rangle$. Specific representation of C_n is given as follows. Operating $\langle n|$ on (3.55) from the left side gives

[5]The concept of the coherent state is first proposed by Schrödinger [3]. Much work on the coherent state in quantum optics was performed by R. J. Glauber. The coherent state is also called the Glauber state in quantum optics.

$$\langle n|\hat{a}|\alpha\rangle = \alpha\langle n|\alpha\rangle. \tag{3.57}$$

On the other hand, from (3.32), $\langle n|\hat{a}$ is

$$\langle n|\hat{a} = \sqrt{n+1}\langle n+1|, \tag{3.58}$$

then, we get

$$\langle n|\hat{a}|\alpha\rangle = \sqrt{n+1}\langle n+1|\alpha\rangle. \tag{3.59}$$

Comparing (3.57) and (3.59), we get

$$\langle n+1|\alpha\rangle = \frac{\alpha}{\sqrt{n+1}}\langle n|\alpha\rangle. \tag{3.60}$$

Applying this relationship recursively, C_n is obtained as

$$\begin{aligned} C_n = \langle n|\alpha\rangle &= \frac{\alpha}{\sqrt{n}}\langle n-1|\alpha\rangle \\ &= \frac{\alpha^2}{\sqrt{n(n-1)}}\langle n-2|\alpha\rangle \cdots = \frac{\alpha^n}{\sqrt{n!}}\langle 0|\alpha\rangle. \end{aligned} \tag{3.61}$$

Then, we get

$$|\alpha\rangle = \langle 0|\alpha\rangle \sum_n \frac{\alpha^n}{\sqrt{n!}}|n\rangle, \tag{3.62}$$

because $\langle 0|\alpha\rangle$ is a constant. Normalization of $|\alpha\rangle$

$$|\alpha\rangle = |\langle 0|\alpha\rangle|^2 \sum_n \frac{|\alpha|^{2n}}{n!} = |\langle 0|\alpha\rangle|^2 e^{|\alpha|^2} = 1 \tag{3.63}$$

requires

$$|\langle 0|\alpha\rangle|^2 = e^{-|\alpha|^2}. \tag{3.64}$$

Finally, we get the expansion form of $|\alpha\rangle$ as

$$|\alpha\rangle = e^{-|\alpha|^2/2} \sum_n \frac{\alpha^n}{\sqrt{n!}}|n\rangle, \tag{3.65}$$

where the global phase is set to be 1. The coherent state is the superposition state of an infinite number of the number states. When we measure the harmonic oscillator in the coherent state, the probability to find it in the eigenstate $|n\rangle$ is

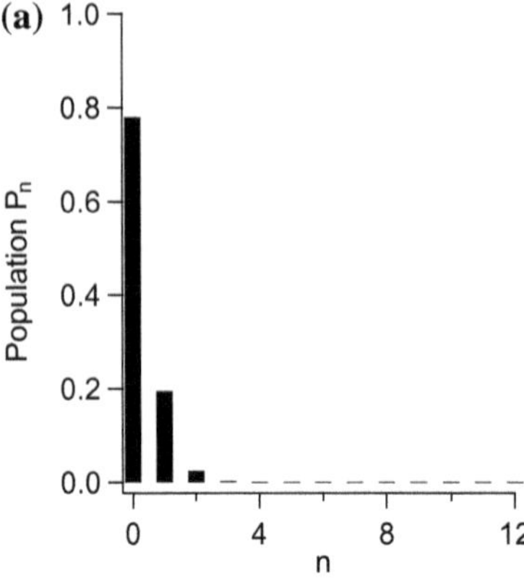

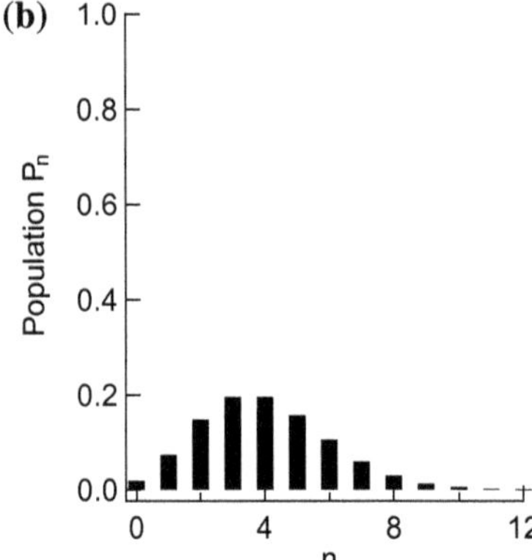

Fig. 3.3 A typical example of the population of number states $|n\rangle$ for the coherent state with $\alpha = 0.5$ (**a**) and $\alpha = 2$ (**b**)

$$P(n) = |\langle n|\alpha\rangle|^2 = e^{-|\alpha|^2}\frac{|\alpha|^{2n}}{n!}, \tag{3.66}$$

which is a Poisson distribution.[6] Then, the the coherent state is the superposition state of energy eigenstates of the harmonic oscillator with a Poisson distribution. Typical examples of the population of the number states in the coherent state is shown in Fig. 3.3.

3.3.2 *Uncertainty Relation of the Coherent State*

Here, we describe the uncertainty relation between position and momentum in the coherent state of the harmonic oscillator. The expectation values for position and

[6]The Poisson distribution $P(n) = e^{-\lambda}\lambda^k/k!$ represents the probability of events occurring k times per unit time, when the events occur randomly λ per unit time.

momentum are

$$\langle x\rangle = \langle\alpha|\hat{x}|\alpha\rangle = \sqrt{\frac{\hbar}{2m\omega}}\langle\alpha|(\hat{a}+\hat{a}^\dagger)|\alpha\rangle = \sqrt{\frac{\hbar}{2m\omega}}(\alpha+\alpha^*)$$
$$\langle p\rangle = \langle\alpha|\hat{p}|\alpha\rangle = -i\sqrt{\frac{m\hbar\omega}{2}}\langle\alpha|(\hat{a}-\hat{a}^\dagger)|\alpha\rangle = -i\sqrt{\frac{m\hbar\omega}{2}}(\alpha-\alpha^*), \quad (3.67)$$

and $\langle x^2\rangle$ and $\langle p^2\rangle$ are

$$\begin{aligned}\langle x^2\rangle &= \langle\alpha|\hat{x}^2|\alpha\rangle = \frac{\hbar}{2m\omega}\langle\alpha|(\hat{a}+\hat{a}^\dagger)(\hat{a}+\hat{a}^\dagger)|\alpha\rangle \\ &= \frac{\hbar}{2m\omega}\langle\alpha|(\hat{a}^2+(\hat{a}^\dagger)^2+2\hat{a}^\dagger\hat{a}+1)|\alpha\rangle = \frac{\hbar}{2m\omega}\left((\alpha+\alpha^*)^2+1\right) \\ \langle p^2\rangle &= \langle\alpha|\hat{p}^2|\alpha\rangle = -\frac{m\hbar\omega}{2}\langle\alpha|(\hat{a}-\hat{a}^\dagger)(\hat{a}-\hat{a}^\dagger)|\alpha\rangle \\ &= -\frac{m\hbar\omega}{2}\left((\alpha-\alpha^*)^2-1\right). \end{aligned} \quad (3.68)$$

Then, we get the variance

$$\begin{aligned}(\Delta x)^2 &= \frac{\hbar}{2m\omega} \\ (\Delta p)^2 &= \frac{m\omega\hbar}{2},\end{aligned} \quad (3.69)$$

and the uncertainty relation is given by

$$\sqrt{(\Delta x)^2(\Delta p)^2} = \frac{\hbar}{2}. \quad (3.70)$$

It shows the coherent state is a minimum uncertainty state for position and momentum.

3.3.3 Displaced Vacuum State

Here, we describe that the coherent state is also derived using a displaced harmonic oscillator. Consider the initial state as the ground state $|0\rangle$ of the harmonic oscillator, in which the potential minimum is at $x = 0$ position (schematically shown in Fig. 3.4). When this harmonic potential is suddenly shifted α along the x-coordinate, the initial state transfers to a new state. This operation is described by a displacement operator $\hat{D}(\alpha)$:

$$\hat{D}(\alpha) = \exp(\alpha\hat{a}^\dagger - \alpha^*\hat{a}). \quad (3.71)$$

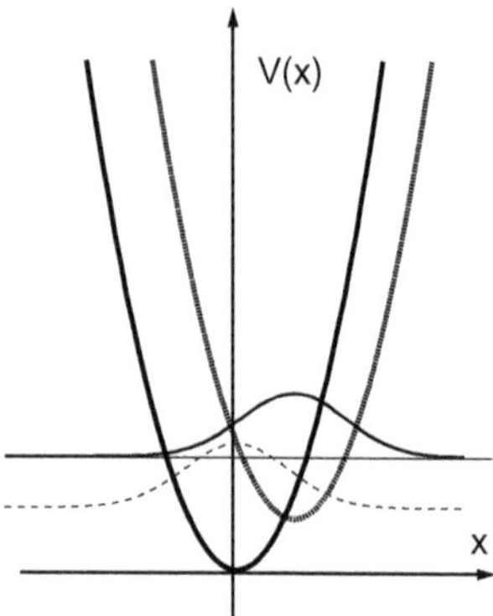

Fig. 3.4 Schematic of the coherent state expressed as the shifted vacuum state. The dotted curve shows a shifted harmonic potential. The dashed Gaussian is the vacuum state

This operator is an unitary operator and satisfies $\hat{D}^\dagger(\alpha) = \hat{D}(-\alpha)$. The displacement operator shifts the annihilation operator by α:

$$\hat{D}^\dagger(\alpha)\hat{a}\hat{D}(\alpha) = \hat{a} + \alpha. \tag{3.72}$$

Operating $\hat{D}(\alpha)$ on $|0\rangle$ gives

$$\begin{aligned}
\hat{D}(\alpha)|0\rangle &= \exp(\alpha\hat{a}^\dagger - \alpha^*\hat{a})|0\rangle \\
&= \exp(\alpha\hat{a}^\dagger)\exp(-\alpha^*\hat{a})\exp\left(-\frac{1}{2}\left[\alpha\hat{a}^\dagger, -\alpha^*\hat{a}\right]\right)|0\rangle \\
&= \exp\left(-\frac{|\alpha|^2}{2}\right)\exp(\alpha\hat{a}^\dagger)\exp(-\alpha^*\hat{a})|0\rangle \\
&= \exp\left(-\frac{|\alpha|^2}{2}\right)\exp(\alpha\hat{a}^\dagger)|0\rangle \\
&= \exp\left(-\frac{|\alpha|^2}{2}\right)\sum_n \frac{\alpha^n}{\sqrt{n!}}|n\rangle.
\end{aligned} \tag{3.73}$$

Then, the suddenly displaced vacuum state is the coherent state.

3.3.4 *Time Evolution of the Coherent State*

The time-dependent coherent state is expressed by

$$|\alpha(t)\rangle = e^{-i\hat{H}t/\hbar}|\alpha\rangle = \langle 0|\alpha\rangle \sum_{n=0}^{\infty} \frac{\alpha^n}{\sqrt{n!}} e^{-i\hat{H}t/\hbar}|n\rangle$$

$$
\begin{aligned}
&= \langle 0|\alpha\rangle \sum_{n=0}^{\infty} \frac{\alpha^n}{\sqrt{n!}} e^{-i(n+1/2)\omega t/\hbar}|n\rangle \\
&= \langle 0|\alpha\rangle e^{-i\omega t/2} \sum_{n=0}^{\infty} \frac{(\alpha e^{-i\omega t}\hat{a}^{\dagger})^n}{n!}|0\rangle \\
&= e^{-|\alpha|^2/2} e^{-i\omega t/2} \exp\left(\alpha e^{-i\omega t}\hat{a}^{\dagger}\right)|0\rangle . \qquad (3.74)
\end{aligned}
$$

Comparing this equation with the time-independent coherent state expanded in the number states

$$
|\alpha\rangle = e^{-|\alpha|^2/2} \exp\left(\alpha \hat{a}^{\dagger}\right)|0\rangle , \qquad (3.75)
$$

we know that α is replaced by $\alpha e^{-i\omega t}$ and the phase factor $e^{-i\omega t/2}$ is multiplied in the time-dependent coherent state. Then, we can rewrite as

$$
|\alpha(t)\rangle = e^{-i\omega t/2}|\alpha e^{-i\omega t}\rangle . \qquad (3.76)
$$

The expectation values of position and momentum are obtained by replacing α to $\alpha e^{-i\omega t}$ in (3.67):

$$
\begin{aligned}
\langle x(t)\rangle &= \sqrt{\frac{2\hbar}{m\omega}}\alpha \cos(\omega t) \\
\langle p(t)\rangle &= = \sqrt{2m\hbar\omega}\alpha \sin(\omega t), \qquad (3.77)
\end{aligned}
$$

for a real α value. Then, the expectation value of position oscillates with the angular frequency ω. As shown in (3.69), the variances of both position and momentum do not depend on α and are time independent. This implies that the shape of the wave packet of the coherent state does not change in time. Then, the coherent state is similar to a classical harmonic oscillator.

3.4 Squeezed States

3.4.1 Squeezed Vacuum States

The coherent state is a minimum uncertainty state for position and momentum ($\Delta x \Delta p = \hbar/2$). In a squeezed state, one of these two variances is reduced while maintaining the minimum uncertainty state [4]. We define new operators ($\hat{b}$ and $\hat{b}^{\dagger}$) using Bogoliubov transformation to the annihilation and creation operators ($\hat{a}$ and $\hat{a}^{\dagger}$) as

$$
\begin{aligned}
\hat{b} &= \mu\hat{a} + \nu\hat{a}^{\dagger} = (\cosh\gamma)\hat{a} + (\sinh\gamma)\hat{a}^{\dagger} \\
\hat{b}^{\dagger} &= \mu\hat{a}^{\dagger} + \nu\hat{a} = (\cosh\gamma)\hat{a}^{\dagger} + (\sinh\gamma)\hat{a}, \qquad (3.78)
\end{aligned}
$$

where μ and ν are hyperbolic functions of a constant γ: $\mu = \cosh\gamma$, $\nu = \sinh\gamma$. Because of $\mu^2 - \nu^2 = 1$, the commutation between $\hat{b}$ and $\hat{b}^\dagger$ is given by

$$[\hat{b}, \hat{b}^\dagger] = \mu^2 - \nu^2 = 1. \tag{3.79}$$

Using the new operators, $\hat{x}$ and $\hat{p}$ are expressed as

$$\begin{aligned} \hat{x} &= \sqrt{\frac{\hbar}{2m\omega}}\left(\hat{a}^\dagger + \hat{a}\right) = e^{-\gamma}\sqrt{\frac{\hbar}{2m\omega}}\left(\hat{b}^\dagger + \hat{b}\right) \\ \hat{p} &= i\sqrt{\frac{m\hbar\omega}{2}}\left(\hat{a}^\dagger - \hat{a}\right) = ie^{\gamma}\sqrt{\frac{m\hbar\omega}{2}}\left(\hat{b}^\dagger - \hat{b}\right). \end{aligned} \tag{3.80}$$

Using e^γ defined by

$$e^\gamma \equiv \sqrt{\frac{m'\omega'}{m\omega}}, \tag{3.81}$$

$\hat{x}$ and $\hat{p}$ are represented by

$$\begin{aligned} \hat{x} &= \sqrt{\frac{\hbar}{2m'\omega'}}\left(\hat{b}^\dagger + \hat{b}\right) \\ \hat{p} &= i\sqrt{\frac{m'\hbar\omega'}{2}}\left(\hat{b}^\dagger - \hat{b}\right). \end{aligned} \tag{3.82}$$

Therefore, $\hat{b}$ and $\hat{b}^\dagger$ are considered to be annihilation and creation operators for the harmonic oscillator ($\hat{H}_b = \hbar\omega'(\hat{b}^\dagger\hat{b} + 1/2)$) defined by the mass m' and the angular frequency of ω' [5]. A sharpened potential for the harmonic oscillator is schematically shown in Fig. 3.5. We set the ground state (vacuum state) of this Hamiltonian as $|0\rangle_b$, which satisfies $\hat{b}|0\rangle_b = 0$. At this state, Δx and Δp are obtained as

$$\begin{aligned} \Delta x &= \sqrt{\frac{\hbar}{2m'\omega'}} = \sqrt{\frac{\hbar}{2m\omega}}e^{-\gamma} \\ \Delta p &= \sqrt{\frac{\hbar m'\omega'}{2}} = \sqrt{\frac{\hbar m\omega}{2}}e^{\gamma}. \end{aligned} \tag{3.83}$$

This state is the squeezed vacuum state, which is also a minimum uncertainty state ($\Delta x \Delta p = \hbar/2$) and Δx is reduced compared to that of the ground state in the original harmonic oscillator with (m, ω) for a positive γ value.

$|0\rangle_b$ is expanded with number states of $\hat{H}$:

$$|0\rangle_b = \sum_{n=0}^{\infty} C_n|n\rangle. \tag{3.84}$$

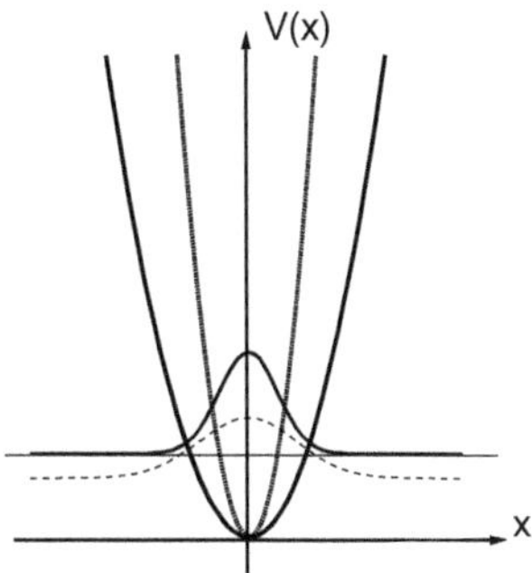

Fig. 3.5 Schematic of the squeezed vacuum state. The dotted curve shows a sharpened harmonic potential. The dashed Gaussian is the vacuum state

Inserting this to $\hat{b}|0\rangle_b = 0$ gives

$$\sum_{n=0}^{\infty} C_n \left(\mu\hat{a} + \nu\hat{a}^\dagger\right)|n\rangle = \sum_{n=0}^{\infty} C_n \left(\mu\sqrt{n}|n-1\rangle + \nu\sqrt{n+1}|n+1\rangle\right)$$
$$= \sum_{n=0}^{\infty} \left(C_{n+1}\mu\sqrt{n+1} + C_{n-1}\nu\sqrt{n}\right)|n\rangle = 0. \quad (3.85)$$

Then, the coefficient of $|n\rangle$ should be zero and we get the recursion relation of

$$C_{n+1} = -\frac{\nu}{\mu}\sqrt{\frac{n}{n+1}}C_{n-1}. \quad (3.86)$$

The $|0\rangle_b$ contains only even or odd number states. We choose the even solution because the ground state $|0\rangle$ is included. When we start from C_0, C_{2m} is obtained by

$$\begin{aligned}
C_2 &= \left(\frac{-\nu}{\mu}\right)\sqrt{\frac{1}{2}}C_0, \\
C_4 &= \left(\frac{-\nu}{\mu}\right)\sqrt{\frac{3}{4}}C_2 = \left(\frac{-\nu}{\mu}\right)^2\sqrt{\frac{3\cdot 1}{4\cdot 2}}C_0 \\
C_6 &= \left(\frac{-\nu}{\mu}\right)\sqrt{\frac{5}{6}}C_4 = \left(\frac{-\nu}{\mu}\right)^3\sqrt{\frac{5\cdot 3\cdot 1}{6\cdot 4\cdot 2}}C_0 \\
&\vdots \\
C_{2m} &= \left(\frac{-\nu}{\mu}\right)\sqrt{\frac{2m-1}{2m}}C_{2(m-1)} = \left(\frac{-\nu}{\mu}\right)^m\sqrt{\frac{(2m-1)!!}{2m!!}}C_0 \\
&= (-1)^m(\tanh\gamma)^{2m}\sqrt{\frac{(2m-1)!!}{(2m)!!}}C_0,
\end{aligned} \quad (3.87)$$

where $(2m)!!$ is the double factorial function $((2m)!! = 2m \cdot (2m-1) \cdot (2m-2) \cdots 0)$ and m is an integer. C_0 is obtained by normalizing the $|0\rangle_b$ as

$$||0\rangle_b|^2 = \sum_{m=0}^{\infty} |C_{2m}|^2 = 1, \tag{3.88}$$

which leads to

$$|C_0|^2 \left(1 + \sum_{m=1}^{\infty} \frac{(\tanh\gamma)^{2m}(2m-1)!!}{(2m)!!}\right) = 1. \tag{3.89}$$

If we use the Taylor expansion of the function

$$f(z) = (1-z)^{-\frac{1}{2}} = 1 + \sum_{n=1}^{\infty} \frac{(2n-1)!!}{2^n n!} z^n, \tag{3.90}$$

we get

$$1 + \sum_{m=1}^{\infty} \frac{(2m-1)!!}{(2m)!!} (\tanh^2\gamma)^m = \left(1 - \tanh^2\gamma\right)^{-\frac{1}{2}}, \tag{3.91}$$

where $(2m)!! = 2^m m!$. The coefficient $|C_0|$ is obtained by

$$|C_0| = \sqrt{\left(1 - \tanh^2\gamma\right)^{\frac{1}{2}}} = \frac{1}{\sqrt{\cosh\gamma}}. \tag{3.92}$$

Finally, $|0\rangle_b$ is expressed by

$$\begin{aligned} |0\rangle_b &= \frac{1}{\sqrt{\cosh\gamma}} \sum_{m=1}^{\infty} (-1)^m \sqrt{\frac{(2m-1)!!}{(2m)!!}} (\tanh\gamma)^m |2m\rangle \\ &= \frac{1}{\sqrt{\cosh\gamma}} \sum_{m=1}^{\infty} (-1)^m \frac{\sqrt{(2m)!}}{2^m m!} (\tanh\gamma)^m |2m\rangle, \end{aligned} \tag{3.93}$$

where $(2m-1)!! = (2m)!/(2^m m!)$. The population of the number state is

$$\begin{aligned} P_{2m} &= |\langle 2m|0\rangle_b|^2 = \frac{(2m)!}{2^{2m}(2m!)^2} \frac{(\tanh\gamma)^{2m}}{\cosh\gamma} \\ P_{2m+1} &= |\langle 2m+1|0\rangle_b|^2 = 0. \end{aligned} \tag{3.94}$$

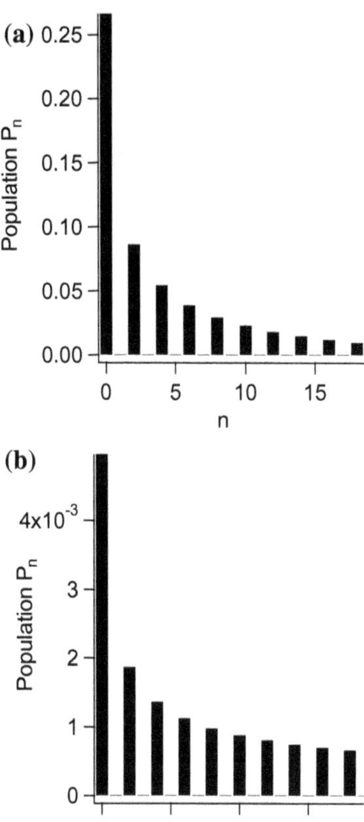

Fig. 3.6 Typical examples of the population of number states $|m\rangle$ for the squeezed vacuum state with $\gamma = 2$ (**a**) and $\gamma = 6$ (**b**)

The squeezed vacuum state $|0\rangle_b$ consists of only even number states. A typical example of the population of the number states in the squeezed states is shown in Fig. 3.6.

Next, we consider time evolution of the squeezed vacuum state. Since the time evolution of the even number states are expressed by $\exp(-i\omega(2m + 1/2)t)|2m\rangle$, the time-dependent squeezed vacuum state $|0(t)\rangle_b$ is

$$|0(t)\rangle_b = \frac{1}{\sqrt{\cosh\gamma}} \sum_{m=1}^{\infty} (-1)^m \frac{\sqrt{(2m)!}}{2^m m!} (\tanh\gamma e^{-i2\omega t})^m |2m\rangle, \tag{3.95}$$

where the phase $\exp(-i\omega t/2)$ is omitted. This means that $\tanh\gamma$ should be replaced by $\tanh\gamma e^{-i2\omega t}$. Then, we need to set $\sinh\gamma \rightarrow \sinh\gamma e^{-i2\omega t}$, which corresponds to $\nu = e^{-i2\omega t}\sinh\gamma$. Here, we introduce a squeeze operator $\hat{S}(\xi)$ as

$$\hat{S}(\xi) \equiv \exp\left(\frac{1}{2}\left(\xi^* \hat{a}^2 - \xi \hat{a}^{\dagger 2}\right)\right), \tag{3.96}$$

where $\xi = \gamma e^{i\theta}$, and set the squeezed vacuum state $|\xi\rangle$ as

$$|\xi\rangle = \hat{S}(\xi)|0\rangle. \tag{3.97}$$

The $\hat{S}(\xi)$ is a unitary operator and $\hat{S}^{\dagger}(\xi) = \hat{S}^{-1}(\xi) = \hat{S}(-\xi)$. By using the Baker–Hausdrof lemma

$$e^{i\lambda\hat{A}}\hat{B}e^{-i\lambda\hat{A}} = \hat{B} + i\lambda[\hat{A}, \hat{B}] + \frac{(i\lambda)^2}{2!}[\hat{A}, [\hat{A}, \hat{B}]] + \cdots, \tag{3.98}$$

we calculate $\hat{S}^{\dagger}(\xi)\hat{a}\hat{S}(\xi)$. When we consider $\hat{A}$ and $\hat{B}$ as

$$\begin{aligned} \hat{A} &= \frac{i}{\lambda}\frac{1}{2}\left(\xi^*\hat{a}^2 - \xi\hat{a}^{\dagger 2}\right) \\ \hat{B} &= \hat{a}, \end{aligned} \tag{3.99}$$

we get

$$\begin{aligned} [\hat{A}, \hat{B}] &= \left(\frac{i}{2\lambda}\right)[(\xi^*\hat{a}^2 - \xi\hat{a}^{\dagger 2}), \hat{a}] \\ &= \left(\frac{i}{2\lambda}\right)(-\xi)[\hat{a}^{\dagger 2}, \hat{a}] = \left(\frac{i}{2\lambda}\right)(2\xi)\hat{a}^{\dagger}. \end{aligned} \tag{3.100}$$

Inserting this result, we get

$$\begin{aligned} [\hat{A}, [\hat{A}, \hat{B}]] &= \left(\frac{i}{2\lambda}\right)^2 (2\xi)[(\xi^*\hat{a}^2 - \xi\hat{a}^{\dagger 2}), \hat{a}^{\dagger}] \\ &= \left(\frac{i}{2\lambda}\right)^2 (2\xi)\xi^*[\hat{a}^2, \hat{a}^{\dagger}] = \left(\frac{i}{2\lambda}\right)^2 (2\xi)(2)\xi^*\hat{a}. \end{aligned} \tag{3.101}$$

The operators $\hat{a}$ and $\hat{a}^{\dagger}$ appear alternately in commutators and we get

$$\begin{aligned} \hat{S}^{\dagger}(\xi)\hat{a}\hat{S}(\xi) &= \hat{a} + (i\lambda)\left(\frac{i}{2\lambda}\right)(2\xi)\hat{a}^{\dagger} + \frac{(i\lambda)^2}{2!}\left(\frac{i}{2\lambda}\right)^2(2|\xi|)^2\hat{a} + \cdots \\ &= \hat{a} - \xi\hat{a}^{\dagger} + \frac{|\xi|^2}{2!}\hat{a} - \frac{\xi|\xi|^2}{3!}\hat{a}^{\dagger}\cdots \\ &= \hat{a}\cosh\gamma - \hat{a}^{\dagger}e^{i\theta}\sinh\gamma = \hat{a}\mu - \hat{a}^{\dagger}\nu, \end{aligned} \tag{3.102}$$

which is the same of the Bogoliubov transformation (3.78) by replacing γ by $-\gamma$. This confirms that $\hat{S}(\xi)$ is the operator which transfers the vacuum state $|0\rangle$ to the squeezed vacuum state.

Next, we calculate expectation values of creation and annihilation operators for the squeezed vacuum state in order to evaluate the expectation value and variance of position ($\hat{x}$) and momentum ($\hat{p}$). At first we calculate the expectation values for the creation and annihilation operators.

$$
\begin{aligned}
\langle\xi|\hat{a}|\xi\rangle &= \langle 0|\hat{S}^\dagger(\xi)\hat{a}\hat{S}(\xi)|0\rangle \\
&= \langle 0|\hat{a}\cosh\gamma - \hat{a}^\dagger e^{i\theta}\sinh\gamma|0\rangle = 0, && (3.103)\\
\langle\xi|\hat{a}^\dagger|\xi\rangle &= \langle 0|\hat{S}^\dagger(\xi)\hat{a}^\dagger\hat{S}(\xi)|0\rangle \\
&= \langle 0|\hat{a}^\dagger\cosh\gamma - \hat{a}e^{-i\theta}\sinh\gamma|0\rangle = 0, && (3.104)\\
\langle\xi|\hat{a}^\dagger\hat{a}|\xi\rangle &= \langle 0|\hat{S}^\dagger(\xi)\hat{a}^\dagger\hat{S}(\xi)\hat{S}^\dagger(\xi)\hat{a}\hat{S}(\xi)|0\rangle \\
&= \langle 0|(\hat{a}^\dagger\cosh\gamma - \hat{a}e^{-i\theta}\sinh\gamma)(\hat{a}\cosh\gamma - \hat{a}^\dagger e^{i\theta}\sinh\gamma)|0\rangle \\
&= \sinh^2\gamma, && (3.105)\\
\langle\xi|\hat{a}\hat{a}^\dagger|\xi\rangle &= \langle 0|\hat{S}^\dagger(\xi)\hat{a}\hat{S}(\xi)\hat{S}^\dagger(\xi)\hat{a}^\dagger\hat{S}(\xi)|0\rangle \\
&= \langle 0|(\hat{a}r\cosh\gamma - \hat{a}^\dagger e^{i\theta}\sinh\gamma)(\hat{a}^\dagger\cosh\gamma - \hat{a}e^{-i\theta}\sinh\gamma)|0\rangle \\
&= \cosh^2\gamma, && (3.106)\\
\langle\xi|\hat{a}\hat{a}|\xi\rangle &= \langle 0|\hat{S}^\dagger(\xi)\hat{a}\hat{S}(\xi)\hat{S}^\dagger(\xi)\hat{a}\hat{S}(\xi)|0\rangle \\
&= \langle 0|(\hat{a}r\cosh\gamma - \hat{a}^\dagger e^{i\theta}\sinh\gamma)(\hat{a}r\cosh\gamma - \hat{a}^\dagger e^{i\theta}\sinh\gamma)|0\rangle \\
&= -e^{i\theta}\sinh\gamma\cosh^2\gamma, && (3.107)\\
\langle\xi|\hat{a}^\dagger\hat{a}^\dagger|\xi\rangle &= \langle 0|\hat{S}^\dagger(\xi)\hat{a}^\dagger\hat{S}(\xi)\hat{S}^\dagger(\xi)\hat{a}^\dagger\hat{S}(\xi)|0\rangle \\
&= \langle 0|(\hat{a}^\dagger\cosh\gamma - \hat{a}e^{-i\theta}\sinh\gamma)(\hat{a}^\dagger\cosh\gamma - \hat{a}e^{-i\theta}\sinh\gamma)|0\rangle \\
&= -e^{-i\theta}\sinh\gamma\cosh^2\gamma. && (3.108)
\end{aligned}
$$

Using above equations, we calculate $\langle\xi|\hat{x}|\xi\rangle$, $\langle\xi|\hat{p}|\xi\rangle$, $\langle\xi|\hat{x}^2|\xi\rangle$, and $\langle\xi|\hat{p}^2|\xi\rangle$:

$$
\begin{aligned}
\langle\xi|\hat{x}|\xi\rangle &= \sqrt{\frac{\hbar}{2m\omega}}\langle\xi|(\hat{a}+\hat{a}^\dagger)|\xi\rangle = 0, && (3.109)\\
\langle\xi|\hat{p}|\xi\rangle &= i\sqrt{\frac{m\hbar\omega}{2}}\langle\xi|(-\hat{a}+\hat{a}^\dagger)|\xi\rangle = 0, && (3.110)\\
\langle\xi|\hat{x}^2|\xi\rangle &= \frac{\hbar}{2m\omega}\langle\xi|(\hat{a}^2+\hat{a}\hat{a}^\dagger+\hat{a}^\dagger\hat{a}+\hat{a}^{\dagger 2})|\xi\rangle \\
&= \frac{\hbar}{2m\omega}\left(\cosh^2\gamma+\sinh^2\gamma-2\cos\theta\cosh\gamma\sinh\gamma\right), && (3.111)\\
\langle\xi|\hat{p}^2|\xi\rangle &= -\frac{m\hbar\omega}{2}\langle\xi|(\hat{a}^2-\hat{a}\hat{a}^\dagger-\hat{a}^\dagger\hat{a}+\hat{a}^{\dagger 2})|\xi\rangle \\
&= \frac{m\hbar\omega}{2}\langle\xi|\left(\cosh^2\gamma+\sinh^2\gamma+2\cos\theta\cosh\gamma\sinh\gamma\right). && (3.112)
\end{aligned}
$$

Then, the expectation values of position and momentum are zero, which are the same as the vacuum state. The variance of position ($\hat{x}$) is the same value of the expectation value of $\hat{x}^2$. For $\theta = 0$, we get

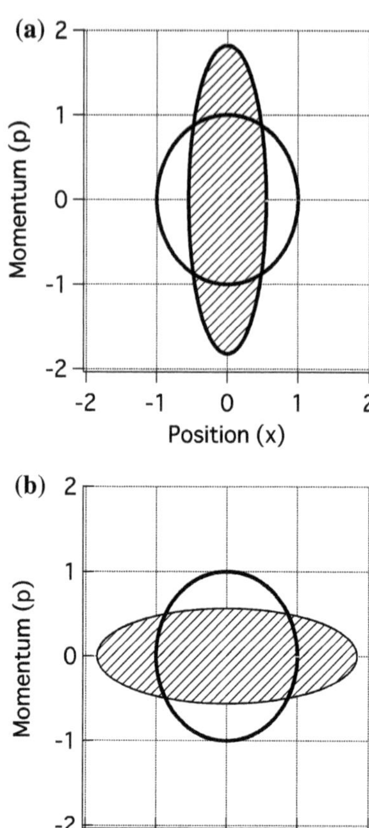

Fig. 3.7 Schematic drawing of the error ellipse of the squeezed vacuum state with $\gamma = 0.6$. x is in a unit of $\sqrt{\hbar/(2m\omega)}$ and y is in a unit of $\sqrt{m\hbar\omega/2}$. **a** $\theta = 0$ and **b** $\theta = \pi$. The center circle represents the error circle of the vacuum state

$$(\Delta x)^2 = \frac{\hbar}{2m\omega}(\cosh\gamma - \sinh\gamma)^2 = \frac{\hbar}{2m\omega}e^{-2\gamma}$$
$$(\Delta p)^2 = \frac{m\hbar\omega}{2}(\cosh\gamma + \sinh\gamma)^2 = \frac{m\hbar\omega}{2}e^{2\gamma}, \tag{3.113}$$

and squeezing exists in the $\hat{x}$ quadrature for $\gamma > 0$. On the other hand, for $\theta = \pi$, squeezing exists in the $\hat{p}$ quadrature. Figure 3.7 shows the error ellipse for the squeezed states.

For an arbitrary angle θ, we use a rotational transformation to operators ($\hat{x}$ and $\hat{p}$) and new operators ($\hat{x}_2$ and $\hat{p}_2$):

$$\begin{pmatrix} \hat{x}_2 \\ \hat{p}_2 \end{pmatrix} = \begin{pmatrix} \cos\frac{\theta}{2} & \sin\frac{\theta}{2} \\ -\sin\frac{\theta}{2} & \cos\frac{\theta}{2} \end{pmatrix} \begin{pmatrix} \hat{x} \\ \hat{p} \end{pmatrix}, \tag{3.114}$$

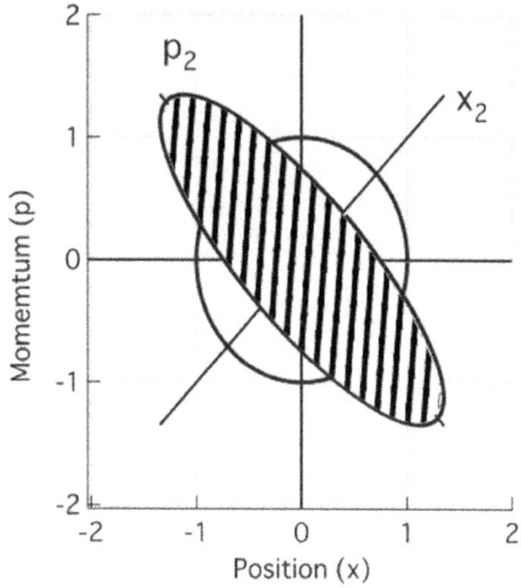

Fig. 3.8 Schematic drawing of the error ellipse of the squeezed vacuum state with $\gamma = 0.6$ and $\theta = \pi/4$. x is in a unit of $\sqrt{\hbar/(2m\omega)}$ and y is in a unit of $\sqrt{m\hbar\omega/2}$. The center circle represents the error circle of the vacuum state

which corresponds to[7]

$$\hat{x}_2 + i\hat{p}_2 = \left(\hat{x} + i\hat{p}\right) e^{-i\theta/2}. \tag{3.115}$$

Then, we get $\hat{x}_2 = e^{-i\theta/2}\hat{x}$ and $\hat{p}_2 = e^{-i\theta/2}\hat{p}$, and the creation and annihilation operators for the new coordinates ($\hat{a}_2$ and $\hat{a}_2^\dagger$) are defined by $\hat{a}_2 = e^{-i\theta/2}\hat{a}$ and $\hat{a}_2^\dagger = e^{-i\theta/2}\hat{a}^\dagger$.

The squeeze operator is expressed using $\hat{a}_2$ and $\hat{a}_2^\dagger$ as

$$\begin{aligned}\hat{S}(\xi) = \hat{S}(re^{i\theta}) &= \exp\left(\frac{1}{2}\left(re^{-i\theta}\hat{a}^2 - re^{i\theta}\hat{a}^{\dagger 2}\right)\right)\\ &= \exp\left(\frac{1}{2}\left(r(e^{-i\theta/2}\hat{a})^2 - r(e^{i\theta}\hat{a}^\dagger)^2\right)\right)\\ &= \exp\left(\frac{1}{2}\left(r\hat{a}_2^2 - r\hat{a}_2^{\dagger 2}\right)\right).\end{aligned} \tag{3.116}$$

This means that the squeeze operator acts with $\theta = 0$ to $\hat{x}_2$ and $\hat{p}_2$. Then, the variance is obtained as

$$\begin{aligned}(\Delta x_2)^2 &= \frac{\hbar}{2m\omega}e^{-2\gamma}\\ (\Delta p_2)^2 &= \frac{m\hbar\omega}{2}e^{2\gamma},\end{aligned} \tag{3.117}$$

[7]This is because $(\hat{x} + i\hat{p})\exp(-i\theta/2) = (\hat{x} + i\hat{p})(\cos(\theta/2) - i\sin(\theta/2)) = (\cos(\theta/2)\hat{x} + \sin(\theta/2)\hat{p}) + i(-\sin(\theta/2)\hat{x} + \cos(\theta/2)\hat{p})$.

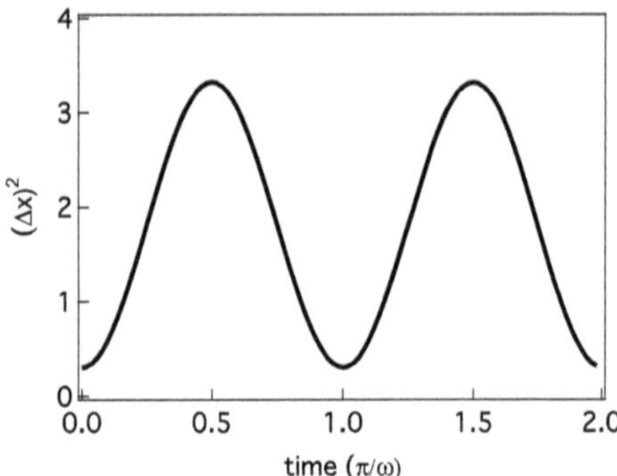

Fig. 3.9 Time dependence of variance ($(\Delta x)^2$) in the vacuum squeezed state with $\gamma = 0.6$

and the squeeze occurs in the x_2 direction. Figure 3.8 shows the error ellipse at $\theta = \pi/4$.

Time evolution of variance is calculated using (3.113) and replacing θ by $2\omega t$:

$$(\Delta x)^2 = \langle \xi | \hat{x}^2 | \xi \rangle = \frac{\hbar}{2m\omega}\left(\cosh^2\gamma + \sinh^2\gamma - 2\cos(2\omega t)\cosh\gamma\sinh\gamma\right)$$
$$(\Delta p)^2 = \langle \xi | \hat{p}^2 | \xi \rangle = \frac{m\hbar\omega}{2}\left(\cosh^2\gamma + \sinh^2\gamma + 2\cos(2\omega t)\cosh\gamma\sinh\gamma\right). \tag{3.118}$$

Then, the variance in the squeezed vacuum state oscillates with the angular frequency of 2ω as shown in Fig. 3.9, though the variance is constant for the coherent state.

3.4.2 *More General Squeezed States*

More general squeezed states are obtained by

$$|\alpha, \xi\rangle = \hat{D}(\alpha)|\xi\rangle = \hat{D}(\alpha)\hat{S}(\xi)|0\rangle. \tag{3.119}$$

The expectation value of the annihilation operator for the state is calculated as

$$\langle \alpha, \xi | \hat{a} | \alpha, \xi \rangle = \langle \xi | \hat{D}^\dagger(\alpha)\hat{a}\hat{D}(\alpha) | \xi \rangle = \langle \xi | (\hat{a} + \alpha) | \xi \rangle$$
$$= \alpha, \tag{3.120}$$

where we used (3.72) and (3.103). A similar calculation gives

$$\langle \alpha, \xi | \hat{a}^\dagger | \alpha, \xi \rangle = \alpha^*$$
$$\langle \alpha, \xi | \hat{a}^2 | \alpha, \xi \rangle = \alpha^2 - e^{i\theta}\sinh\gamma\cosh^2\gamma = \alpha^2 + \langle \xi | \hat{a}^2 | \xi \rangle$$
$$\langle \alpha, \xi | \hat{a}^{\dagger 2} | \alpha, \xi \rangle = \alpha^{*2} - e^{-i\theta}\sinh\gamma\cosh^2\gamma = \alpha^{*2} + \langle \xi | \hat{a}^{\dagger 2} | \xi \rangle$$

$$\langle \alpha, \xi | \hat{a}^\dagger \hat{a} | \alpha, \xi \rangle = |\alpha|^2 + \sinh^2 = |\alpha|^2 + \langle \xi | \hat{a}^\dagger \hat{a} | \xi \rangle$$
$$\langle \alpha, \xi | \hat{a} \hat{a}^\dagger | \alpha, \xi \rangle = |\alpha|^2 + \cosh^2 \gamma = |\alpha|^2 + \langle \xi | \hat{a} \hat{a}^\dagger | \xi \rangle. \tag{3.121}$$

Using these equations, we get the expectation value of the position and the momentum as

$$\begin{aligned} \langle \hat{x} \rangle = \langle \alpha, \xi | \hat{x} | \alpha, \xi \rangle &= \sqrt{\frac{\hbar}{2m\omega}} \langle \alpha, \xi | (\hat{a} + \hat{a}^\dagger) | \alpha, \xi \rangle \\ &= \sqrt{\frac{\hbar}{2m\omega}} (\alpha + \alpha^*) \\ \langle \hat{p} \rangle = \langle \alpha, \xi | \hat{p} | \alpha, \xi \rangle &= -i \sqrt{\frac{m\hbar\omega}{2}} \langle \alpha, \xi | (\hat{a} - \hat{a}^\dagger) | \alpha, \xi \rangle \\ &= -i \sqrt{\frac{m\hbar\omega}{2}} (\alpha - \alpha^*), \end{aligned} \tag{3.122}$$

which are the same as those of the coherent states (3.67). Similarly, we get

$$\begin{aligned} \langle \hat{x}^2 \rangle &= \frac{\hbar}{2m\omega} \langle \alpha, \xi | (\hat{a}\hat{a} + \hat{a}^\dagger \hat{a} + \hat{a}\hat{a}^\dagger + \hat{a}^\dagger \hat{a}^\dagger) | \alpha, \xi \rangle \\ &= \langle \xi | \hat{x}^2 | \xi \rangle + \frac{\hbar}{2m\omega} (\alpha^2 + \alpha^{*2} + 2|\alpha|^2) \\ \langle \hat{p}^2 \rangle &= -\frac{m\hbar\omega}{2} \langle \alpha, \xi | (\hat{a}\hat{a} + \hat{a}^\dagger \hat{a} - \hat{a}\hat{a}^\dagger - \hat{a}^\dagger \hat{a}^\dagger) | \alpha, \xi \rangle \\ &= \langle \xi | \hat{p}^2 | \xi \rangle - \frac{m\hbar\omega}{2} (\alpha^2 + \alpha^{*2} - 2|\alpha|^2). \end{aligned} \tag{3.123}$$

The variances are obtained as

$$\begin{aligned} (\Delta x)^2 &= \langle \hat{x}^2 \rangle - (\langle \hat{x} \rangle)^2 = \langle \xi | \hat{x}^2 | \xi \rangle \\ (\Delta p)^2 &= \langle \hat{p}^2 \rangle - (\langle \hat{p} \rangle)^2 = \langle \xi | \hat{p}^2 | \xi \rangle, \end{aligned} \tag{3.124}$$

and the same as those of the squeezed vacuum state (3.113). Therefore, in the general squeezed state $|\alpha, \xi\rangle$, the mean value of the position is the same as that of the coherent state $|\alpha\rangle$ and the variance is the same as that of the squeezed vacuum state $|\xi\rangle$. The time evolution of the position $\langle \hat{x} \rangle$ and the variance $(\Delta x)^2$ oscillates with the frequency of ω and 2ω, respectively.

3.5 Summary

The one-dimensional harmonic oscillator was investigated quantum mechanically. The Hamiltonian of the harmonic oscillator was expressed using the creation $\hat{a}$ and annihilation $\hat{a}^\dagger$ operators. The energy eigenstate is expressed using the number state

$|n\rangle$. The coherent state, which is a minimum uncertainty state, is introduced. The squeezed state, in which the variance of the position and the momentum is reduced while maintaining the minimum uncertainty state, is also introduced. In the coherent state, the mean value of the position oscillates with the angular frequency ω with keeping its variance constant. In the vacuum squeezed state, the mean value of the position is constant and its variance oscillates with 2ω [6].

References

1. Sakurai, J.J.: Modern Quantum Mechanics, Revised edn. Addison-Wesley Publishing Company Inc., Reading (1994)
2. Takahashi, Y.: Ba no Ryoshiron I (in Japanese: Quantum Field Theory I). Baifukan, Tokyo (1974)
3. Schrödinger, E.: Der steige Ubergang von der Mikro- zur Makromechanik. Die Nautureissenschaften **14**, 664–666 (1926)
4. Gerry, C.C., Knight, P.L.: Introductory Quantum Optics. Cambridge University Press, New York (2008)
5. Zwiebach, B.: Quantum dynamics (2013). https://ocw.mit.edu/courses/physics/8-05-quantum-physics-ii-fall-2013/lecture-notes/MIT8_05F13_Chap_06.pdf
6. I wrote this chapter referencing Refs. [1–5] and following books: Yoshio Kuramoto and Junichi Ezawa, *Ryoushi Rikigaku (in Japanese, Quantum Mechanics)*, Asakura Shyoten (2008); Keiji Igi and Hikari Kawai, *Kiso Ryousi Rikigaku (in Japanese, Fundamental Quantum Mechanics)*, Kodansya (2007); Masahito Ueda *Gendai Ryoushi Buturigaku (in Japanese, Modern Quantum Physics)*, Baifukan (2004); Akira Shimizu, *Shin-han Rryoushi Ron no Kiso (in Japanese, New edition Fundamental of Quantum Physics)*, Science Shya (2004); Kyo Inoue, *Kogaku Kei no Tameno Ryoushi Kogaku (in Japanese, Quantum Optics for Engineer)*, Morikita Shyoten (2015); Masahiro Matsuoka, *Ryoushi Kogaku (in Japanese, Quantum Optics)*, Shokabou (2000)

Chapter 4
Lattice Vibration and Phonon

A phonon is a quantum description of lattice vibrations in solids. This chapter summaries basics of lattice vibration and phonons using a linear atomic chain. At first, we calculate dynamics of atomic motions with classical mechanics and introduce a plane wave expansion to express collective atomic motions. A phonon is introduced by using field quantization. Optical and acoustic phonons are explained by using a linear diatomic chain.

4.1 Linear Atomic Chain

We consider a simple linear atomic chain as a model of one-dimensional lattice vibration. At first we treat it in classical mechanics and then in quantum mechanics.

4.1.1 Classical Treatment

Suppose atoms with mass of m stay in line with spacing a.[1] The displacement of the lth atom from its equilibrium position is set to be q^l as shown in Fig. 4.1. The spring constant between atoms is f. The equation of motion of this atom is

$$m\ddot{q}^l = f(q^{l+1} + q^{l-1} - 2q^l). \tag{4.1}$$

We adapt a periodic boundary condition ($q^l = q^{l+N}$) for a total of N atoms. Using a plane wave expansion, the displacement is expressed by

[1] Here, we referred to the method described in [1] for classical and quantum mechanics of a linear atomic chain.

K. Nakamura, *Quantum Phononics*, Springer Tracts in Modern Physics 282,
https://doi.org/10.1007/978-3-030-11924-9_4

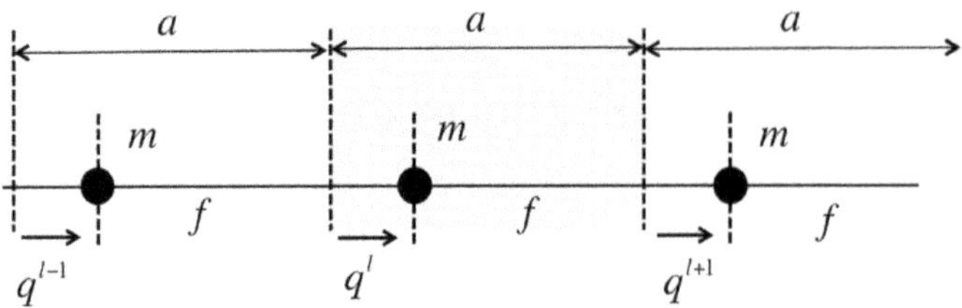

Fig. 4.1 Schematic of the linear atomic chain. The unit cell with the lattice constant a contains one atom with mass of m. The force constant is f, and q^l is atomic displacement in the lth unit cell

$$q^l(t) = \frac{1}{\sqrt{N}} e^{ikla} Q_k(t), \tag{4.2}$$

where k is the wave vector which satisfies

$$k = \frac{2n\pi}{Na}. \tag{4.3}$$

n is an integer in a range of $(-N/2, N/2)$. $1/\sqrt{N}$ is the normalization factor, which satisfies

$$\sum_{l=1}^{N} \left(\frac{1}{\sqrt{N}} e^{ikla} \right)^* \left(\frac{1}{\sqrt{N}} e^{ik'la} \right) = \delta_{k,k'}. \tag{4.4}$$

Using this plane wave expansion, we get

$$\ddot{Q}_k(t) = D \left(e^{ika} + e^{-ika} - 2 \right) Q_k(t), \tag{4.5}$$

where $D = f/m$. This is the equation of simple harmonic oscillation for Q_k, and then we express Q_k as

$$Q_k(t) = e^{-i\omega_k t} A_k, \tag{4.6}$$

and get the equation

$$-\omega_k^2 e^{-i\omega_k t} A_k = D \left(e^{ika} + e^{-ika} - 2 \right) e^{-i\omega_k t} A_k. \tag{4.7}$$

Then we get a dispersion relation between the frequency (ω) and the wave vector (k) as

$$\omega_k = 2\sqrt{D} |\sin(ka/2)|. \tag{4.8}$$

The dispersion curve is shown in Fig. 4.2. The frequency ω_k is equal to ω_{-k}.

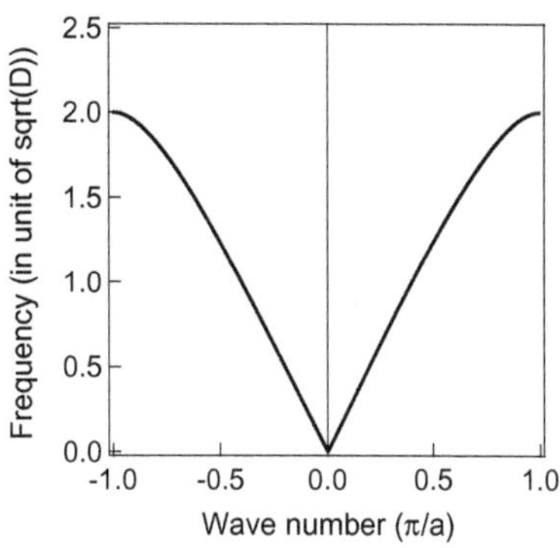

Fig. 4.2 Phonon dispersion curve of the linear atomic chain

The displacement of the lth atom is expressed as

$$\begin{aligned} q^l(t) &= \sum_k \left(\frac{1}{\sqrt{N}} e^{ikla} e^{-i\omega_k t} A_k + \frac{1}{\sqrt{N}} e^{-ikla} e^{i\omega_k t} A_k^* \right) \\ &= \sum_k \left(\frac{1}{\sqrt{N}} e^{ikla} Q_k(t) + \frac{1}{\sqrt{N}} e^{-ikla} Q_k^*(t) \right). \end{aligned} \tag{4.9}$$

The Hamiltonian H of the lattice vibration is obtained by

$$H = \sum_l \frac{p^l(t)^2}{2m} + \frac{f}{2} \sum_l \left(q^l(t) - q^{l+1}(t) \right)^2, \tag{4.10}$$

where $p^l(t)$ is the momentum

$$\begin{aligned} p^l(t) = m\dot{q}^l(t) &= \sum_k \left(m \frac{1}{\sqrt{N}} e^{ikla} \dot{Q}_k(t) + c.c. \right) \\ &= \sum_k \left(-i\omega_k m \frac{1}{\sqrt{N}} e^{ikla} Q_k(t) + c.c. \right). \end{aligned} \tag{4.11}$$

Here $c.c.$ means the complex conjugate. The kinetic energy T is obtained by

$$\begin{aligned} T = \frac{p^l(t)^2}{2m} &= \frac{1}{2m} \sum_l \left(\sum_k -i\omega_k m \frac{1}{\sqrt{N}} e^{ikla} Q_k(t) + c.c. \right) \\ &\times \left(\sum_{k'} -i\omega_{k'} m \frac{1}{\sqrt{N}} e^{ik'la} Q_{k'}(t) + c.c. \right). \end{aligned} \tag{4.12}$$

Using the orthogonal condition

$$\frac{1}{N}\sum_{l} e^{ikla+ik'la} = \delta_{k,-k}, \tag{4.13}$$

we rewrite the kinetic energy as

$$\begin{aligned} T &= \frac{m}{2}\sum_{k}\Big(-\omega_k\omega_{-k}Q_k(t)Q_{-k}(t) - \omega_k\omega_{-k}Q_k^*(t)Q_{-k}^*(t) \\ &\quad +\omega_k^2 Q_k(t)Q_k^*(t) + \omega_k^2 Q_k^*(t)Q_k(t)\Big) \\ &= \frac{m}{2}\sum_{k}\omega_k^2\Big(-Q_k(t)Q_{-k}(t) - Q_k^*(t)Q_{-k}^*(t) \\ &\quad +Q_k(t)Q_k^*(t) + Q_k^*(t)Q_k(t)\Big). \end{aligned} \tag{4.14}$$

The potential energy V is obtained by

$$\begin{aligned} V &= \frac{f}{2}\sum_{l}\Big((q^l(t))^2 + (q^{l+1}(t))^2 - 2q^l(t)q^{l+1}(t)\Big) \\ &= \frac{f}{2}\sum_{l}\Big(2(q^l(t))^2 - q^{l-1}(t)q^l(t) - q^{l+1}(t)q^l(t)\Big), \end{aligned} \tag{4.15}$$

by using different pairs of displacements. Each term of the right-hand side in the equation is expressed as V_1, V_2, and V_3. V_1 is obtained by

$$\begin{aligned} V_1 &= f\sum_{l}\Bigg(\sum_{k}\Big(\frac{1}{\sqrt{N}}e^{ikla}Q_k(t) + \frac{1}{\sqrt{N}}e^{-ikla}Q_k^*(t)\Big) \\ &\quad \times\sum_{k'}\Big(\frac{1}{\sqrt{N}}e^{ik'la}Q_{k'}(t) + \frac{1}{\sqrt{N}}e^{-ik'la}Q_{k'}^*(t)\Big)\Bigg) \\ &= f\sum_{k}\Big(Q_k(t)Q_{-k}(t) + Q_k(t)Q_k^*(t) \\ &\quad +Q_k^*(t)Q_k(t) + Q_k^*(t)Q_{-k}^*(t)\Big). \end{aligned} \tag{4.16}$$

A similar calculation gives V_2 and V_3 as

$$V_2 = \frac{-f}{2}\sum_{k}\Big(e^{-ila}Q_k(t)Q_{-k}(t) + e^{-ika}Q_k(t)Q_k^*(t)$$

$$+e^{ika} Q_k^*(t) Q_k(t) + e^{ika} Q_k^*(t) Q_{-k}^*(t) \Big)$$

$$V_3 = \frac{-f}{2} \sum_k \Big(e^{ila} Q_k(t) Q_{-k}(t) + e^{ika} Q_k(t) Q_k^*(t)$$

$$+e^{-ika} Q_k^*(t) Q_k(t) + e^{-ika} Q_k^*(t) Q_{-k}^*(t) \Big). \tag{4.17}$$

Using the dispersion relation

$$f - \frac{f}{2} e^{-ika} - \frac{f}{2} e^{ika} = \frac{m\omega_k^2}{2}, \tag{4.18}$$

we get

$$V = \frac{m\omega_k^2}{2} \left(Q_k(t) Q_{-k}(t) + Q_k(t) Q_k^*(t) + Q_k^*(t) Q_k(t) + Q_k^*(t) Q_{-k}^*(t) \right). \tag{4.19}$$

Then we get the Hamiltonian as

$$H = T + V = \sum_k m\omega_k^2 (Q_k^*(t) Q_k(t) + Q_k(t) Q_k^*(t)). \tag{4.20}$$

By introducing the dimensionless parameter b_k, we can simplify the formula as

$$H = \sum_k \frac{\hbar\omega_k}{2} (b_k^* b_k + b_k b_k^*), \tag{4.21}$$

where the relation between b_k and Q_k is

$$b_k = \sqrt{\frac{2m\omega_k}{\hbar}} Q_k(t). \tag{4.22}$$

The lattice vibration is expressed as the summation of the harmonic oscillators with their coordinates $Q_k(t)$, which are not the same as the atomic displacements $q^l(t)$.[2]

4.1.2 Quantum Treatment

For the canonical quantization, the coordinate q_l and momentum p_l are replaced by linear operators $\hat{q}_l$ and $\hat{p}_l$, which satisfy the commutation relations

[2]At the Γ-point ($k = 0$), the atomic displacement in the lth unit cell is the same of $Q_k(t)/\sqrt{N}$.

$$\begin{aligned} \left[\hat{q}^l, \hat{p}^n\right] &= i\hbar\delta_{l,n}, \\ \left[\hat{q}^l, \hat{q}^n\right] &= \left[\hat{p}^l, \hat{p}^n\right] = 0. \end{aligned} \tag{4.23}$$

In the Schrödinger picture, a dynamical variable is expressed by a time-independent operator, where we omit time dependence in variables and operators. The amplitudes of lattice vibration Q_k and Q_k^* are also replaced by $\hat{Q}_k$ and $\hat{Q}_k^\dagger$ and the $\hat{q}^l$ are expressed by

$$\begin{aligned} \hat{q}^l &= \sum_k \left(\frac{1}{\sqrt{N}} e^{ikla} \hat{Q}_k + \frac{1}{\sqrt{N}} e^{-ikla} \hat{Q}_k^\dagger \right) \\ &= \sum_k \frac{1}{\sqrt{N}} e^{ikla} \left(\hat{Q}_k + \hat{Q}_{-k}^\dagger \right), \end{aligned} \tag{4.24}$$

because the summations of $\sum_k$ and $\sum_{-k}$ are the same for the periodic boundary conditions. The Fourier transformation of $\hat{q}^l$ is expressed by

$$\frac{1}{\sqrt{N}} \sum_{l=1}^{N} e^{-ikla} \hat{q}^l = \hat{Q}_k + \hat{Q}_{-k}^\dagger. \tag{4.25}$$

Here we introduce a new operator $\hat{\alpha}_k$ as

$$\hat{\alpha}_k \equiv \hat{Q}_k + \hat{Q}_{-k}^\dagger. \tag{4.26}$$

Applying the same procedure to the momentum operator $\hat{p}_l$ gives

$$\begin{aligned} \hat{p}^l &= \sum_k \left(-i\omega_k m \frac{1}{\sqrt{N}} e^{ikla} \hat{Q}_k + i\omega_k m \frac{1}{\sqrt{N}} e^{-ikla} \hat{Q}_k^\dagger \right), \\ &= \frac{m}{\sqrt{N}} \sum_k e^{ikla} \left(-i\omega_k \hat{Q}_k + i\omega_{-k} \hat{Q}_{-k}^\dagger \right). \end{aligned} \tag{4.27}$$

Its Fourier transform is

$$\begin{aligned} \frac{1}{\sqrt{N}} \sum_{l=1}^{N} e^{-ikla} \hat{p}^l &= -i\omega_k m \hat{Q}_k + i\omega_{-k} m \hat{Q}_{-k}^\dagger \\ &= \hat{\beta}_k m \omega_k, \end{aligned} \tag{4.28}$$

where we used $\omega_k = \omega_{-k}$ and introduced a new operator $\hat{\beta}$

$$\hat{\beta}_k \equiv -i\hat{Q}_k + i\hat{Q}_{-k}^\dagger. \tag{4.29}$$

Using (4.26) and (4.29), we get

$$\hat{Q}_k = \frac{1}{2}\left(\hat{\alpha}_k + i\hat{\beta}_k\right),$$
$$\hat{Q}^\dagger_{-k} = \frac{1}{2}\left(\hat{\alpha}_k - i\hat{\beta}_k\right). \tag{4.30}$$

The commutation relation between $\hat{Q}_k$ and $\hat{Q}^\dagger_{k'}$ at the same time is calculated by

$$\begin{aligned}[\hat{Q}_k, \hat{Q}^\dagger_{k'}] &= \hat{Q}_k\hat{Q}^\dagger_{k'} - \hat{Q}^\dagger_{k'}\hat{Q}_k \\ &= \frac{1}{4}\Big((\hat{\alpha}_k + i\hat{\beta}_k)(\hat{\alpha}_{-k'} - i\hat{\beta}_{-k'}) - (\hat{\alpha}_{-k'} - i\hat{\beta}_{-k'})(\hat{\alpha}_k + i\hat{\beta}_k)\Big) \\ &= \frac{1}{4}\Big([\hat{\alpha}_k, \hat{\alpha}_{-k'}] + [\hat{\beta}_k, \hat{\beta}_{-k'}] + i[\hat{\beta}_k, \hat{\alpha}_{-k'}] - i[\hat{\alpha}_k, \hat{\beta}_{-k'}]\Big). \end{aligned} \tag{4.31}$$

The operators $\hat{\alpha}_k$ and $\hat{\alpha}^\dagger_{k'}$ and $\hat{\beta}_k$ and $\hat{\beta}^\dagger_{k'}$ are commutative, because $\hat{\alpha}_k$ and $\hat{\beta}_k$ include only $\hat{q}^l$ or $\hat{p}^l$, respectively: $\left[\hat{\alpha}_k, \hat{\alpha}_{-k'}\right] = 0$, $\left[\hat{\beta}_k, \hat{\beta}_{-k'}\right] = 0$. Then we get

$$[\hat{Q}_k, \hat{Q}^\dagger_{k'}] = \frac{1}{4}\Big(i[\hat{\beta}_{-k'}, \hat{\alpha}_k] + i[\hat{\beta}_k, \hat{\alpha}_{-k'}]\Big). \tag{4.32}$$

We rewrite this commutation relation using $\hat{q}^l$ and $\hat{p}^l$ and get

$$\begin{aligned}[\hat{Q}_k, \hat{Q}^\dagger_{k'}] = \frac{i}{4Nm}\bigg(&\frac{1}{\omega_{k'}}\sum_{l'=1}^{N} e^{ik'l'a}\sum_{l=1}^{N} e^{-ikla}[\hat{p}^{l'}, \hat{q}^l] \\ &+ \frac{1}{\omega_k}\sum_{l'=1}^{N} e^{-ikl'a}\sum_{l=1}^{N} e^{ik'la}[\hat{p}^{l'}, \hat{q}^l]\bigg). \end{aligned} \tag{4.33}$$

Using the commutation relation $[\hat{p}^{l'}, \hat{q}^l] = -i\hbar\delta_{l,l'}$ we get

$$[\hat{Q}_k, \hat{Q}^\dagger_{k'}] = \frac{\hbar}{4Nm}\left(\sum_{l=1}^{N}\left(\frac{1}{\omega_{k'}} + \frac{1}{\omega_{k'}}\right)e^{ia(k'-k)}\right) = \frac{\hbar}{2m\omega_k}\delta_{k,k'}. \tag{4.34}$$

We introduce a dimensionless operator $\hat{b}_k$ as

$$\hat{b}_k \equiv \sqrt{\frac{2m\omega_k}{\hbar}}\hat{Q}_k, \tag{4.35}$$

where this operator satisfies the commutation relation

$$[\hat{b}_k, \hat{b}^\dagger_{k'}] = \delta_{k,k'}. \tag{4.36}$$

Finally, the Hamiltonian is obtained by

$$\hat{H} = \sum_k \hbar\omega_k \left(\hat{b}_k^\dagger \hat{b}_k + \frac{1}{2} \right). \tag{4.37}$$

This Hamiltonian has the same form as the harmonic oscillator, and the operators $\hat{b}_k^\dagger$ and $\hat{b}_k$ are creation and annihilation operators, respectively. This quantum of the lattice vibration is called a phonon.[3] The phonon is a boson like the photon. In the stationary condition, the lattice wave with the wave number k is explained to be occupied by n_k phonons.

4.2 Linear Diatomic Chain

4.2.1 Classical Treatment

In the linear atomic chain, there is only one phonon mode (acoustic phonon). Next, we consider a linear diatomic chain, which includes two atoms (atom 1 and atom 2) in a unit cell.[4] The atom 1 and atom 2 have masses of m_1 and m_2, respectively. a is a length of the unit cell. In the linear diatomic chain, another phonon mode (optical phonon) exists in addition to the acoustic phonon. The atomic displacement of each atom in the lth unit cell is expressed by q_κ^l for atom κ, $\kappa = 1, 2$. By adapting the periodic boundary condition for N unit cells, $q_\kappa^{l+N} = q_\kappa^l$ (Fig. 4.3).

The kinetic energy T and the potential energy V are obtained by

$$\begin{aligned} T &= \sum_{l,\kappa} \frac{(p_\kappa^l)^2}{2m_\kappa} \\ V &= \frac{1}{2} \sum_l \left(f \left(q_2^l - q_1^l\right)^2 + g \left(q_1^{l+1} - q_2^l\right)^2 \right), \end{aligned} \tag{4.38}$$

where p_κ^l is a momentum of the atom κ and defined by

$$p_\kappa^l = m_\kappa \dot{q}_\kappa^l = m_\kappa \frac{\mathrm{d}}{\mathrm{d}t} q_\kappa^l. \tag{4.39}$$

[3]A phonon is a collective oscillation, and is sometimes referred to as a quasi particle. The quasi particle is a fictitious body consisting of the original real individual particle pulse train of disturbed neighbors [2].

[4]We referred to the method described in [3] for classical and quantum mechanics of the linear diatomic chain.

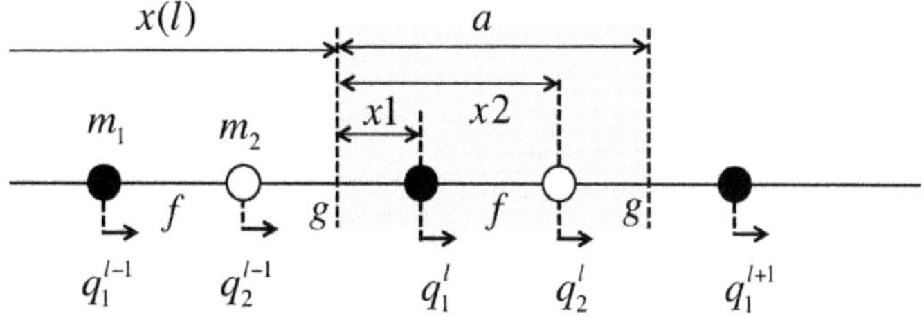

Fig. 4.3 Schematic of the linear diatomic chain. The unit cell contains two atoms (atom 1 and atom 2 represented by black and white balls) with masses of m_1 and m_2, respectively. The force constants between two atom 1 and atom 2 are f and g. $x(l)$ represents the position of the lth unit cell. $x1$ and $x2$ represent the equilibrium position of the atom 1 and atom 2 from the edge of the unit cell, respectively

For the atom 1 in the lth unit cell, the equation of motion is

$$\begin{aligned} m_1 \frac{\mathrm{d}^2}{\mathrm{d}t^2} q_1^l &= -\frac{\mathrm{d}V}{\mathrm{d}q_1^l} \\ &= -\frac{1}{2}\frac{\mathrm{d}}{\mathrm{d}q_1^l}\left(f\left(q_2^l - q_1^l\right)^2 + g\left(q_1^{l+1} - q_2^l\right)^2 \right) \\ &= f\left(q_2^l - q_1^l\right) + g\left(q_2^{l-1} - q_1^l\right), \end{aligned} \tag{4.40}$$

where $-\mathrm{d}V/\mathrm{d}q_1^l$ is the force acting on the atom 1. The equation for atom 2 is calculated in a similar way:

$$m_2 \frac{\mathrm{d}^2}{\mathrm{d}t^2} q_1^l = -\frac{\mathrm{d}V}{\mathrm{d}q_2^l} = f\left(q_1^l - q_2^l\right) + g\left(q_1^{l+1} - q_2^l\right). \tag{4.41}$$

Using the plane wave expansion, displacements are expressed by

$$q_\kappa^l = \frac{1}{2}\frac{1}{\sqrt{N m_\kappa}}\left(A_0(k) a_\kappa(k) e^{ikx_\kappa^l - i\omega(k)t} \right) + c.c., \tag{4.42}$$

where k is the wave number, $\omega(k)$ is the vibration frequency, and x_κ^l is the equilibrium position of the atom κ in the lth unit cell. Inserting (4.42) into (4.40), we get

$$\begin{aligned} -(\omega(k))^2 \sqrt{m_1} a_1(k) e^{ikx_1^l} &= f\left(\frac{1}{\sqrt{m_2}} a_2(k) e^{ikx_2^l} - \frac{1}{\sqrt{m_1}} a_1(k) e^{ikx_1^l} \right) \\ &\quad + g\left(\frac{1}{\sqrt{m_2}} a_2(k) e^{ikx_2^l} - \frac{1}{\sqrt{m_1}} a_1(k) e^{ikx_1^l} \right), \end{aligned} \tag{4.43}$$

where we neglected the complex conjugate term. The equilibrium position of each atom has relationships as follows: $x_1^l = x(l) + x1$, $x_2^l = x(l) + x2$, and $x_2^{l-1} =$

$x(l) + x2 - a$. Using these relations, removing the factor $e^{ikx(l)}$ gives

$$
\begin{aligned}
-(\omega(k))^2\sqrt{m_1}a_1(k)e^{ikx1} &= f\left(\frac{1}{\sqrt{m_2}}a_2(k)e^{ikx2} - \frac{1}{\sqrt{m_1}}a_1(k)e^{ikx1}\right) \\
&\quad + g\left(\frac{1}{\sqrt{m_2}}a_2(k)e^{-ika}e^{ikx2} - \frac{1}{\sqrt{m_1}}a_1(k)e^{ikx1}\right). \qquad (4.44)
\end{aligned}
$$

Rearranging the equation gives

$$
(\omega(k))^2 a_1(k) = \frac{f+g}{m_1}a_1(k) - \frac{f+ge^{-ika}}{\sqrt{m_1 m_2}}e^{ik(x2-x1)}a_2(k). \qquad (4.45)
$$

A similar calculation for $a_2(k)$ gives

$$
(\omega(k))^2 a_2(k) = \frac{f+g}{m_2}a_2(k) - \frac{f+ge^{ika}}{\sqrt{m_1 m_2}}e^{-ik(x2-x1)}a_1(k). \qquad (4.46)
$$

These equations are expressed by using a matrix

$$
\begin{pmatrix} D_{11} - \omega(k)^2 & D_{12} \\ D_{21} & D_{22} - \omega(k)^2 \end{pmatrix}\begin{pmatrix} a_1(k) \\ a_2(k) \end{pmatrix} = 0, \qquad (4.47)
$$

where

$$
\begin{aligned}
D_{11} &= \frac{f+g}{m_1} \\
D_{12} &= \frac{f+ge^{-ika}}{\sqrt{m_1 m_2}}e^{ik(x2-x1)} \\
D_{22} &= \frac{f+g}{m_2} \\
D_{21} &= \frac{f+ge^{ika}}{\sqrt{m_1 m_2}}e^{-ik(x2-x1)}.
\end{aligned} \qquad (4.48)
$$

$a_1(k)$ and $a_2(k)$ have nonzero solutions when $\omega(k)$ satisfies

$$
\begin{vmatrix} D_{11} - \omega(k)^2 & D_{12} \\ D_{21} & D_{22} - \omega(k)^2 \end{vmatrix} = 0, \qquad (4.49)
$$

then

$$
\begin{aligned}
(D_{11} - \omega(k)^2)(D_{22} - \omega(k)^2) - D_{12}D_{21} &= 0, \\
\omega(k)^4 - (D_{11} + D_{22})\omega(k)^2 + D_{11}D_{22} - D_{12}D_{21} &= 0.
\end{aligned} \qquad (4.50)
$$

Here we calculate

$$
\begin{aligned}
D_{11} + D_{22} &= \frac{f+g}{m_1} + \frac{f+g}{m_2} = (f+g)\frac{m_1+m_2}{m_1 m_2} \\
&= \frac{f+g}{\mu}, \\
D_{11}D_{22} - D_{12}D_{21} &= \frac{2fg}{m_1 m_2}(1-\cos(ka)) = \frac{4fg}{m_1 m_2}\sin^2\left(\frac{ka}{2}\right),
\end{aligned}
\tag{4.51}
$$

where μ is a reduced mass $\mu \equiv (m_1 m_2)/(m_1 + m_2)$. Then we get the quadratic equation of $\omega^2(k)$ as

$$
\left(\omega^2(k)\right)^2 - \frac{f+g}{\mu}\omega^2(k) + \frac{4fg}{m_1 m_2}\sin^2\left(\frac{ka}{2}\right). \tag{4.52}
$$

The solution of $\omega^2(k)$ is

$$
\omega^2(k) = \frac{f+g}{2\mu}\left(1 \pm \sqrt{1 - \frac{16\mu^2}{m_1 m_2}\frac{fg}{(f+g)^2}\sin^2\left(\frac{ka}{2}\right)}\right). \tag{4.53}
$$

Since frequencies are positive, we get

$$
\omega_j(k) = \sqrt{\frac{f+g}{2\mu}\left(1 \pm \sqrt{1 - \frac{16\mu^2}{m_1 m_2}\frac{fg}{(f+g)^2}\sin^2\left(\frac{ka}{2}\right)}\right)}, \tag{4.54}
$$

where $j = 1, 2$ corresponding to $+$ term and $-$ term, respectively. The corresponding eigenvector is defined by using $a_{\kappa,j}(k)$. The frequency ω_j is called a normal frequency. The lattice vibration with the normal frequency is also called a normal mode. In the normal mode, all atoms oscillate with the same frequency ω_j in phase. In general, the lattice vibration consists of many normal mode oscillations, and then the oscillation of each atom is not in phase.

4.2.1.1 Example of Linear Diatomic Chain

Here, we consider a special case for the linear diatomic chain: $f = g$ for the force constant, $m_2 > m_1$ for the atomic mass and $x_1 = 0, x_2 = a/2$ for the equilibrium atomic position. From (4.54), the eigenfrequency is

$$
\omega_j(k) = \sqrt{\frac{3f}{2m_1}\left(1 \pm \sqrt{1 - \frac{8}{9}\sin^2\left(\frac{ka}{2}\right)}\right)}. \tag{4.55}
$$

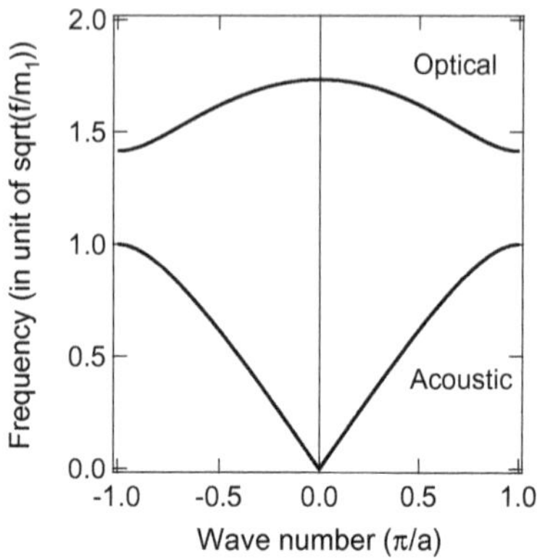

Fig. 4.4 Phonon dispersion curves of the diatomic linear chain with the force constant ($f = g$), the atomic mass ($m_2 = 2m_1$), and lattice constant of a in the first Brillouin zone. The higher frequency mode is an optical mode and the lower one is an acoustic mode

The frequency is plotted as a function of the wave number in Fig. 4.4. There is a gap, in which the frequency does not exist, between the high- and low-frequency branches.

At $ka \simeq 0$ condition (long-wavelength approximation), the eigenfrequencies were obtained to be $\omega_1 = 0$ and $\omega_2 = \sqrt{3f/m_1}$, where we set the low and high frequencies ω_1 and ω_2, respectively. The atomic displacements are calculated by using (4.42) as

$$\frac{q_1^l}{q_2^l} = 1 \tag{4.56}$$

for the $j = 1$ branch and

$$\frac{q_1^l}{q_2^l} = -\frac{m_2}{m_1} = -2 \tag{4.57}$$

for the $j = 2$ branch. All atoms move in the same direction with the same displacement, and the whole chain moves in the $j = 1$ branch, which is called an acoustic mode. In the $j = 2$ branch, two neighboring atoms move in opposite directions to each other and the center of mass does not move. This oscillation is called an optical mode, because it can induce electronic polarization and interacts with the electronic field of light for ions. At around the $ka = \pi$ condition, the eigen frequencies are $\omega_1 = \sqrt{f/m_1}$ and $\omega_2 = \sqrt{2f/m_1}$ for the acoustic and optical modes, respectively. The optical phonons with a wave number of $k \sim 0$ are quite important because they are excited by light, for example, in Raman spectroscopy (Fig. 4.5).

4.2.1.2 Normal Coordinates and Hamiltonian

The displacement of atom κ with the wavenumber k is expressed by using normal coordinates $Q_j(k)$ as

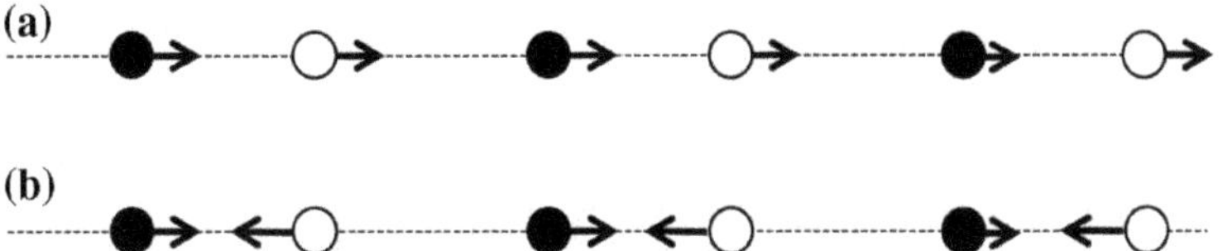

Fig. 4.5 Directions of atomic displacement for the acoustic mode (**a**) and the optical mode (**b**) at $ka \simeq 0$

$$q_\kappa^l = \frac{1}{\sqrt{Nm_\kappa}} \left(\sum_{j,k \geq 0} a_{\kappa,j}(k) e^{ikx_\kappa^l} Q_j(k) + \sum_{j,k>0} a_{\kappa,j}(-k) e^{-ikx_\kappa^l} Q_j(-k) \right), \quad (4.58)$$

where

$$\begin{aligned} Q_j(k) &= \frac{1}{2} \left(A_{0,j}(k) e^{-i\omega_j(k)t} + A_{0,j}(k) e^{i\omega_j(-k)t} \right) \\ &= \frac{1}{2} \left(A_{0,j}(k) e^{-i\omega_j(k)t} + A_{0,j}^*(-k) e^{i\omega_j(k)t} \right). \end{aligned} \quad (4.59)$$

The most important merit to use normal coordinates is that the Hamiltonian is diagonalized:

$$H = T + V = \frac{1}{2} \sum_{k,j} \left\{ P_j(k) P_j^*(k) + Q_j(k) Q_j^*(k) \omega_j^2 \right\}. \quad (4.60)$$

Derivation of this relation is shown in Appendix A. Equations of motion in a Hamiltonian form are

$$\begin{aligned} \dot{Q}_j(k) &= \frac{\partial H}{\partial P_j^*(k)} = P_j(k) \\ \dot{P}_j(k) &= \frac{-\partial H}{\partial Q_j^*(k)}, \end{aligned} \quad (4.61)$$

then we get

$$\ddot{Q}_j(k) = \dot{P}_j(k) = -\omega_j^2 Q_j(k), \quad (4.62)$$

which is an equation of motion for uncoupled 2N harmonic oscillators. In a normal coordinate which belongs to an eigenfrequency defined by the mode and the wave number, all atoms oscillate with the same phase.

4.2.2 Quantum Treatment

Using the same method as the linear atomic chain, we replace $Q_j(k)$ and $P_j(k)$ by $\hat{Q}_j(k)$ and $\hat{P}_j(k)$, which satisfy the commutation relation:

$$\left[\hat{Q}_j^*(k), \hat{P}_{j'}(k')\right] = \left[\hat{Q}_j(k), \hat{P}_{j'}^*(k')\right] = i\hbar\Delta(k-k')\delta_{j,j'}$$
$$\left[\hat{Q}_j^*(k), \hat{Q}_j(k')\right] = \left[\hat{P}_j^*(k), \hat{P}_j(k')\right] = 0, \tag{4.63}$$

where $\Delta(k-k') = 1$ for $k = k' \pm 2\pi n/a$ and $\Delta(k-k') = 0$ for otherwise. In addition, we introduce creation and annihilation operators ($\hat{b}_j^\dagger(k)$ and $\hat{b}_j(k)$) as

$$\hat{b}_j(k) = \sqrt{\frac{\omega_j(k)}{2\hbar}}\hat{Q}_j(k) + i\sqrt{\frac{1}{2\hbar\omega_j(k)}}\hat{P}_j(k)$$
$$\hat{b}_j^\dagger(k) = \sqrt{\frac{\omega_j(k)}{2\hbar}}\hat{Q}_j(k) - i\sqrt{\frac{1}{2\hbar\omega_j(k)}}\hat{P}_j(k). \tag{4.64}$$

The Hamiltonian is expressed by

$$H = \frac{1}{2}\sum_{k,j}\left(\hat{P}_j^*(k)\hat{P}_j(k) + \omega_j^2\hat{Q}_j^*(k)\hat{Q}_j(k)\right)$$
$$= \frac{1}{2}\sum_{k,j}\hbar\omega_j\left(\hat{b}_j^\dagger(k)\hat{b}_j(k) + \frac{1}{2}\right). \tag{4.65}$$

Therefore, the system is calculated as uncoupled 2N harmonic oscillators.

4.3 Summary

A phonon is a quantum description of lattice vibrations in solids. At first, we calculate dynamics of atomic motions with classical mechanics and introduce a plane wave expansion to express collective atomic motions. A phonon is introduced by using field quantization. The phonon in a normal mode is expressed by the Hamiltonian which has the same form used for the harmonic oscillator using the creation and annihilation operators for the normal mode. Optical and acoustic phonons are explained by using a linear diatomic chain.

References

1. Haken, H.: Quantum Filed Theory of Solids. North-Holland Publishing Company, Amsterdam (1976)
2. Bruesch, P.: Phonons: Theory and Experiments I. Springer Series in Solid-State Sciences 34. Springer, NewYork (1982)
3. Mattuck, R.D.: A Guide to Feynman Diagrams in the Many-Body Problems, 2nd edn. Dover Publications Inc., New York (1992)

Chapter 5
Coherent Phonons: Experiment

Coherent phonons are coherently excited by an ultrashort optical pulse and used to study dynamics of optical phonons. In this chapter, we present a brief history of experiments and phenomenological theories of coherent phonons. A typical example of the experimental technique and results of coherent phonons using a pump and probe protocol is explained. Several selected examples of coherent phonons in semiconductors and semimetals are presented. In addition, the squeezed phonons, which are essential nonclassical states, are described.

5.1 A Brief History

5.1.1 Experiments of Coherent Optical Phonons

When an ultrashort optical pulse with duration much shorter than an oscillation period of optical phonons is irradiated on materials, the optical phonons are excited impulsively and coherently. Oscillation of the induced optical phonons is detected via a transient change in reflectivity or transmissivity. Such impulsively excited phonons are called coherent phonons.[1] Experimental studies of coherent optical phonons started in the mid-1980s. Keith Nelson and his coworkers demonstrated femtosecond time-resolved measurements of optical phonons in the organic molecular crystal (α-perylene) [1]. They used 70 fs optical pulses with wavelength of 620 nm and performed pump–probe experiments. The α-perylene crystal has two pairs of four planar molecules in a unit cell. They observed coherent oscillations due to vibrational (33 and 80 cm^{-1}) and translational (104 cm^{-1}) modes.

[1]It should be mentioned that the coherent phonons are not necessarily a coherent state of the phonon.

K. Nakamura, *Quantum Phononics*, Springer Tracts in Modern Physics 282,
https://doi.org/10.1007/978-3-030-11924-9_5

Cho et al. [2] studied transient reflectivity of a semiconductor crystal (GaAs) and found 8.8-THz coherent phonon oscillations due to the longitudinal optical (LO) phonon in GaAs. Coherent phonons in semimetals (bismuth and antimony) were observed through reflectivity modulation by Cheng et al. [3]. The observed phonon mode was an isotropic optical (A_{1g}) mode with 2.9 and 4.5-THz frequencies for the Bi and Sb samples, respectively. Chwalek et al. [4] demonstrated the time-resolved observation of A_{1g}-mode phonon in the semiconducting cuprate compound $YBa_2Cu_3O_{6+x}$ ($x < 0.4$) using 100 fs pulses.

Coherent phonons have now been extensively studied in various materials, for example, in semimetals [5–7], semiconductors [8–13], superconductors [14–17], oxides [18–20], topological insulators [21–26], and carbon materials [27–30], using pump–probe optical measurements.

5.1.2 Generation Mechanism of Optical Coherent Phonons

Generation mechanisms of the optical coherent phonons are often classified into three mechanisms: the impulsive stimulated Raman scattering (ISRS) [31, 32] for transparent conditions, the displacive excitation of coherent phonons (DECP) [33] for opaque conditions, and the screening of the surface-charge field for a polar semiconductor such as gallium arsenide [2, 34]. The equation of motion for an optical phonon is expressed in a normal coordinate Q using a classical harmonic oscillator model:

$$\frac{\mathrm{d}^2 Q}{\mathrm{d}t^2} + 2\gamma \frac{\mathrm{d}Q}{\mathrm{d}t} + \omega_0^2 Q = F(t), \tag{5.1}$$

where γ and ω_0 are the vibrational damping constant and frequency, respectively, $F(t)$ is the driving force.

The ISRS mechanism was first proposed by Yan et al. [31]. When an ultrashort optical pulse passes through a Raman-active material, phonons, which have vibrational periods longer than the pulse width, are excited by the ISRS process, because the Stokes frequency is contained within the bandwidth of the pulse. The driving force is given by

$$F(t) = \frac{1}{2} N \left(\frac{\partial \alpha}{\partial Q} \right) E^2, \tag{5.2}$$

where N is the number density of oscillators, α is the differential polarizability tensor, and E is the electric field of the pulse. The obtained oscillation of Q is a sine-like oscillation ($Q_0 \sin(\omega_0 t)$), which starts from the equilibrium position.

The DECP mechanism was proposed by Zeiger et al. [33] for coherent optical phonons with A_1 symmetry in semimetal materials and semiconductors. In this mechanism, the origin of the oscillation is a change in the quasi-equilibrium A_1 nuclear coordinate $Q_0(t)$ from the equilibrium position $Q_0 = 0$ before laser irradiation. This shift is due to photoexcited electrons ($n(t)$: electron density per unit volume) in the excited state. They assumed the linear dependence of the shift on the electron density: $Q_0(t) = \kappa n(t)$. The driving force is expressed as

$$\begin{aligned} F(t) &= \omega_0\, 2Q_0(t) = \omega_0^2 \kappa n(t) \\ \frac{\mathrm{d}n(t)}{\mathrm{d}t} &= \rho E_p g(t) - \beta n(t), \end{aligned} \tag{5.3}$$

where E_p and $g(t)$ are the energy and a normalized pulse-shape function of the optical pulse, respectively, ρ is a constant of proportionality for carrier generation, and β is a rate constant for relaxation. The oscillation shows a $\cos(\omega_0 t)$ dependence in DECP mechanism. The DECP mechanism is analogous to the mechanism using a displaced potential for the electronic excited state, which is proposed for coherent oscillations in large organic molecules.

For polar semiconductors, another generation mechanism due to a change in a surface-charge field is proposed [2, 34]. The driving force of this mechanism is the sudden depolarization of the crystal lattice due to the ultrafast change of intrinsic surface-charge field in its depletion layer which is given by

$$F(t) = \int_{\infty}^{t} J_j(t')\mathrm{d}t', \tag{5.4}$$

where $J_j(t)$ is a current associated with drift of photoexcited carriers in the surface-charge field.

Kuznetsov and Stanton [35] proposed a microscopic theory in which the dynamics of electrons and phonons are described by kinetic equations using quantum-mechanical operators. They used the Hamiltonian:

$$\begin{aligned} \hat{H}_{el} &= \sum_{k,\alpha} \varepsilon_{\alpha k} \hat{c}^{\dagger}_{\alpha k} \hat{c}_{\alpha k} + \sum_{q} \hbar\omega_q \hat{b}^{\dagger}_q \hat{b}_q \\ &+ \sum_{\alpha,k,q} M_{kq} \left(\hat{b}_q + \hat{b}^{\dagger}_q\right) \hat{c}^{\dagger}_{\alpha k} \hat{c}_{\alpha k+q}, \end{aligned} \tag{5.5}$$

where $\hat{c}^{\dagger}$, $\hat{c}$ are the electron creation and annihilation operators in k space, respectively, $\varepsilon_{\alpha k}$ is the energy dispersion in band $\alpha = \{c, v\}$ (conduction or valence band). ω_q, $\hat{b}^{\dagger}$, and $\hat{b}$ are the phonon angular frequency, the phonon creation, and annihilation operators. The last term is an interaction term describing electron–phonon coupling, and M_{kq} is the coupling constant related to a deformation potential. They defined a coherent amplitude $D_q \equiv \langle\hat{b}\rangle + \langle\hat{b}^{\dagger}\rangle$ of the qth phonon mode using the statistical average $\langle\hat{b}\rangle$ and $\langle\hat{b}^{\dagger}\rangle$ and obtained the equation of motion using D_q:

$$\frac{\partial^2}{\partial t^2} D_q + \omega_q^2 D_q = -2\omega_q \sum_{\alpha,k} M^{\alpha}_{kq} n^{\alpha}_{k,k+q}, \tag{5.6}$$

where $n^{\alpha}_{k,k+q}$ is the electronic density matrix. The derived equation is similar to the phenomenological equation for classical oscillators.

Numerical simulations for generating coherent phonons have been studied using the time-dependent density functional theory (TDDFT) by Yabana and his coworkers [36, 37]. They used a classical electronic field for the optical pulse, the first principle theory for electronic states using DFT, and classical dynamics for the atomic motion. The ISRS process is confirmed to be a mechanism for generating the driving force in a diamond sample.

5.2 Experiments

The coherent phonons are usually excited by irradiation of an ultrashort optical pulse and detected by using transient optical transmission (or reflection) measurements using a pump–probe technique. Recently, using ultrashort X-ray pulses, the coherent phonon oscillation is also detected via a time-resolved X-ray diffraction technique.

5.2.1 *Optical Measurement*

The coherent optical phonons are excited and measured via pump–probe transmission (or reflection) measurements using femtosecond laser pulses. The pump pulse with duration shorter than a vibration period of a phonon induces the phonons with finite timing. Induced phonons oscillate in phase and cause a modulation in macroscopic electric susceptibility, which can be detected via transmissivity (or reflectivity) of the probe pulse. Figure 5.1 shows a typical example of the pump–probe experiments.

Femtosecond optical pulses with a central wavelength of 800 nm are generated by a Ti:sapphire oscillator, which is excited by 532 nm light from a continuous wave laser. The femtosecond pulse is separated into two pulses (pump and probe pulses) by a beam splitter. The pump and probe pulses pass through different optical paths, which have their lengths controlled. Since the light travels with the constant light speed (2.99×10^8 m/s), the arrival time at the sample position is different for the pump and probe pulses. The difference of 299 nm in the optical paths corresponds to 1 fs. In this setup, path length for the probe pulse is controlled by using a shaker (Scan delay unit), which oscillates at 20 Hz. Intensity (signal intensity: I_s) of the probe pulse transmitted through a sample is detected by a photodiode (PD2), which does not need a high time resolution. The probe pulse intensity (reference intensity: I_r) is monitored by a photodiode (PD1). The differential signal ($I_s - I_r$) is introduced to a current amplifier and detected by an oscilloscope. The pump and probe pulses are focused on the sample by using a lens or a parabolic mirror.[2] The output from the

[2]A parabolic mirror is often used to focus an ultrashort optical pulse shorter than approximately 30 fs in order to avoid strong chirp effects by a lens.

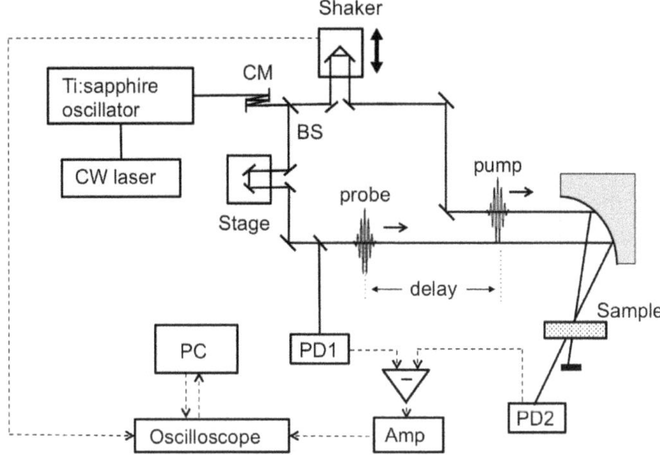

Fig. 5.1 Schematic of the experimental setup for coherent phonon measurements using a pump and probe technique. This is a setup for the transient transmission measurement. BS: beam splitter, CM: chirp mirror, PD: photodiode, Amp: current amplifier, PC: personal computer. Solid and dashed lines represent the optical path and the electric connection, respectively

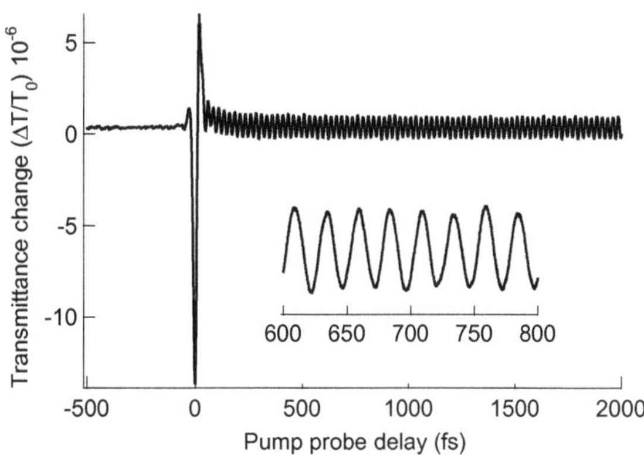

Fig. 5.2 Transient transmittance from a single crystal of diamond measured by using a pump and probe technique. Inset is an enlarged signal in the delay range between 500 and 600 fs. The oscillation period is 25.1 ± 0.03 fs

laser oscillator is reflected by a pair of chirp mirrors in order to compensate for the group velocity dispersion of optics and to become the shortest pulse at the sample position.

Figure 5.2 shows a typical example of the femtosecond time-resolved transmittance measurement of diamond using the pump and probe technique and sub-10 fs laser pulses at room temperatures.[3] There is a strong peak at zero delay due to the non-

[3]The used sample is a single crystal of diamond with a (100) face. The sample was fabricated by chemical vapor deposition and obtained from EPD corporation and has intermediate type between Ib and IIa with a size of 5 mm square and 0.7 mm thickness [38]. The optical pulse is characterized immediately behind the output port of the laser oscillator using a spectrometer and a frequency-resolved autocorrelation (FRAC) technique. The spectrum showed a maximum-intensity wavelength

linear response of diamond by overlapping the pump and probe pulses in the sample. After the strong peak, there is a modulation caused by the coherent optical phonons in diamond. Inset in Fig. 5.2 is the enlarged signal in delay between 600 and 800 fs. The oscillation period is 25.1 ± 0.03 fs (frequency of 39.9 ± 0.05 THz), which is the same value of the optical phonon of diamond at Γ-point ($k = 0$) obtained by Raman spectroscopy.[4] The used laser pulse is near infrared, and its energy (1.5 eV) is well below the band gap (7.3 eV) of diamond. Then, the generation mechanism in this experiment is ISRS. The oscillation is well reproduced by a damped harmonic oscillation

$$\frac{\Delta T(t)}{T_0} = C \exp(-t/\tau) \sin(\omega t + \theta), \tag{5.7}$$

where C is an oscillation amplitude, τ is a coherent time, ω is an angular frequency, and θ is an initial phase. The coherent time is obtained to be approximately 6 ps.

5.2.1.1 Semimetals

Coherent phonons in semimetals were first reported by Cheng et al. [3]. They measured femtosecond time-resolved reflection from bismuth and antimony samples using 70 fs pulses of laser light at 1.98 eV (625.8 nm) and the pump and probe technique. Bismuth and antimony crystallize in the A7 structure, which is a trigonally distorted cubic structure with two atoms per unit cell. There are two optical phonon modes: isotropic A_{1g} and anisotropic E_g modes. The transient reflection intensity showed coherent oscillations with frequencies of 2.9 and 4.5 THz, which correspond to the A_{1g} optical phonon frequencies at Γ-point ($k = 0$) for bismuth and antimony, respectively. The anisotropic E_g mode phonon oscillation was not observed in their time-resolved measurement.

By using an electro-optical (EO) sampling technique [34], the E_g mode phonon oscillation was observed in bismuth [39]. In this technique, the polarization of the probe pulse is set perpendicular to that of the pump pulse. The reflected probe pulse is directed into a polarizing beamsplitter, and the parallel and perpendicular components are detected by using balanced photodetectors. Thus, the difference between two orthogonal components $R_x - R_y$ is detected. A schematic of the EO sampling is shown in Fig. 5.3. Using this technique, the isotropic oscillation (A_{1g}) signal is suppressed and the weak anisotropic oscillation (E_g) signal can be observed [39]. In addition to the A_{1g} coherent phonons, the E_g coherent phonons oscillating with oscillation period of 475 fs (2.1 THz) were observed. The phonon decay time was

of 792 nm with a bandwidth of approximately 200 nm, and the estimated pulsed width was 8.2 fs at full width at half maximum from the FRAC measurement.

[4]Coherent phonons in diamond were first reported by Ishioka et al. [28] by using sub-10 fs, 395 nm pulses via transient reflection measurements.

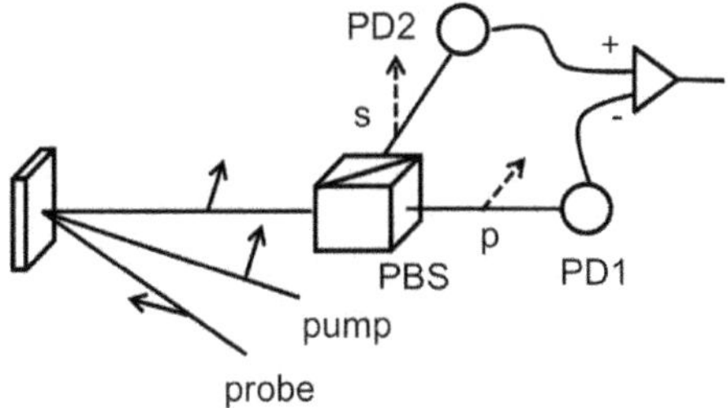

Fig. 5.3 Schematic of reflective electro-optic (EO) sampling technique. The reflected probe pulse is directed to a polarizing beam splitter (PBS), and p- and ps-polarized components are separated and detected by photodiodes (PD1 and PD2). PD1 and PD2 have reverse-bias voltages. An arrow shows a polarization direction of pulses

obtained to be 3.71 and 2.21 ps for A_{1g} and E_g, respectively, by using curve fitting with two damping oscillations.

5.2.1.2 Semiconductors

Transient pump–probe reflectivity measurements have been performed to observe coherent oscillation of longitudinal optical (LO) phonons in GaAs by Cho et al. [2]. The sample used was (100)-oriented intrinsic GaAs (i-GaAs). They used 50-fs pulses at 2 eV and observed a periodic oscillation with an oscillation frequency of 8.8 ± 0.15 THz (oscillation period of ∼114 fs), which corresponds to the LO phonon in i-GaAs. The dephasing time was obtained to be 2.0 ps at excitation density close to 10^{17}cm^{-3} and 0.7 ps at density of 10^{18}cm^{-3} at room temperature. Coherent LO phonons were also observed in n- and p-doped GaAs ($N_D = 2 \times 10^{18}$ cm^{-3}) using the reflective EO sampling [34]. They measured cosine-like oscillation with a frequency of 8.75 THz. The phase of the coherent LO-phonon oscillation in n-doped GaAs was shifted by π from that in p-doped GaAs. The initial phase is determined by the sign of the driving force, which is due to the screening of the surface-space-charge field. In a polar semiconductor, the valence and conduction bands are bent in a depletion layer near the surface due to pining of the Fermi level at the charged surface state. Thus, a macroscopic surface-space-charge field is created perpendicular to the sample surface. When electron–hole pairs or electronic polarization are created by pumping in the depletion layer, the space-charge-field is screened and a sudden change of constraining force causes driving force of the lattice vibration. The bands bend upward and downward at the surface for n- and p-doped GaAs, respectively. Then the driving force in the n- and p-doped GaAs has different directions and the initial phase was shifted by π. In a complementary experiment, Pfeifer et al. [34] applied an external electric field to the sample and found an enhancement of the phonon amplitude. Typical examples of the pump-probe transient reflectivity measurements of GaAs are shown in Fig. 5.4.

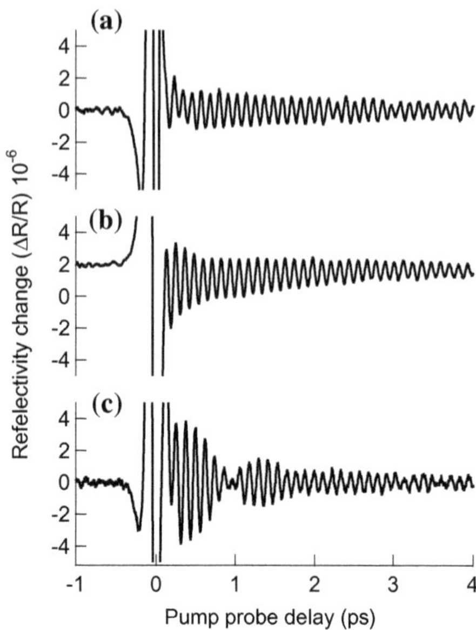

Fig. 5.4 Reflectivity change in the photoexcited intrinsic GaAs (**a**), p-doped GaAs (**b**) and n-doped GaAs (**c**) measured using pump–probe reflective electro-optical sampling. The coherent oscillation due to LO phonons is shown in the intrinsic GaAs. In the n-doped GaAs, there is beating due to two coherent oscillations (LO phonon and LOPC). In the p-doped GaAs, very small oscillation of LOPC overlaps with that of LO phonons

In addition to LO phonons, the LO phonon–plasmon coupled (LOPC) oscillation is excited in n-doped GaAs [40]. The frequency of the plasmon, which is proportional to square root of electron density ($\sqrt{N_e}$), approaches to the LO-phonon frequency at $N_e \sim 10^{18}$ cm^{-3} and the plasmon oscillation couples to LO-phonon oscillation. Hu et al. [41] irradiated a 45-fs pulse at 1.55 eV on the n-doped GaAs (100) with $N_e = 10^{18}$ cm^{-3} sample and excited the total density ($N_e = 4.2 \pm 0.5 \times 10^{18}$ cm^{-3}) of photoexcited and intrinsic carriers. The coherent oscillations of the LO phonon and LOPC are 8.8 and 7.7 THz, respectively, at room temperature. The dephasing time of LO phonon (1.5 ps) is longer than that of LOPC (0.9 ps) and increases as temperature decreases. The temperature dependence of the decay rate $1/\tau$ was well fitted by the Klemens channel

$$\frac{1}{\tau} = \Gamma_0 \left(1 + 2 \frac{1}{\exp(\hbar\omega_{LA}/(2k_B T)) - 1} \right), \tag{5.8}$$

where Γ_0 is a fitting parameter (fitted value is 0.26 ps^{-1}), k_B is the Boltzmann constant, and T is the temperature. In the Klemens channel, an optical phonon decays into two acoustic phonons with a frequency of $\omega_{LA} = \omega_{LO}/2$ and equal but opposite wave vectors. On the other hand, the dephasing time of LOPC was less temperature dependent.

Many experimental examples of the coherent phonons in bulk and low-dimensional semiconductors are reviewed in the literature [9].

5.2.2 X-Ray Diffraction Measurement

Coherent optical phonons have also been detected by using ultrafast time-resolved X-ray diffraction, where coherent phonons are excited by a femtosecond optical pulse and detected by using an ultrashort X-ray pulse [42–46]. The first clear experiments were demonstrated by Sokolowski-Tinten et al. [42] on a bismuth film sample. They performed femtosecond time-resolved X-ray diffraction using a laser pump and X-ray probe technique with 120-fs laser pulses and laser-plasma X-rays (LPX).[5] The X-ray diffraction intensity of the (222) reflection showed oscillation just after the pump-laser pulse irradiation with an oscillation period of 467 fs (frequency of 2.12 THz). This oscillation is assigned to the A_{1g} optical phonon. The observed frequency was downshifted from that of the pristine sample (2.92 THz) because the sample was highly excited by the laser irradiation. The diffraction intensity is dependent on the square modulus of the structure factor $F(h, k, l, t)$, where h, k, and l are Miller indices and t is time. The time dependence of the structure factor is expressed by

$$F(h, k, l, t) = \sum_{j=1}^{N} f_i \exp\left(-i\mathbf{G}_j \cdot \left(\mathbf{r}_j + \boldsymbol{\delta}_j\right)\right), \tag{5.9}$$

where f_j, $\mathbf{r}_j$, and $\boldsymbol{\delta}_j$ are the atomic scattering factor, the crystallographic atomic position, and the atomic-deviation vector of the jth atom in a unit cell. For coherently excited phonons, the atomic-deviation vector is $\boldsymbol{\delta}_j = \mathbf{u}_j \sin(\omega t + \theta)$, where $\mathbf{u}_j$ and θ are the maximum deviation vector and initial phase. Thus, the diffraction intensity oscillates according to the optical phonon oscillation. Unlike the optical reflection or transmission measurements, the value of the atomic displacement is directly determined using the X-ray diffraction experiment. Sokolowski-Tinten et al. [42] reported that the maximum atomic displacement is approximately 5–8% of the nearest-neighbor distance (0.31 nm) between the bismuth atoms in the A7 structure at pump-laser fluence of 6 mJ/cm^2.

5.3 Squeezed Phonons

As discussed in Chap. 3, a squeezed state is an essential nonclassical state. In the squeezed state, it is possible to reduce the fluctuations of one variable to a level lower than that of the vacuum state at the expense of the fluctuations of its conjugate variable. Because of this possibility to overcome the limitation of the uncertainty

---

[5] When the metal target is irradiated by an intense ultrashort laser pulse, with intensity much larger than 10^{16} W/cm^2, the rising edge of the pulse has enough energy to produce plasma. Electrons in the plasma are accelerated more than keV by the main part of the pulse. The high-energy electrons collide with target atom, excite inner shell electrons, and produce characteristic X-rays. For example, short-pulsed K_α and K_β emissions are reported for a copper target for 42-fs laser-irradiation experiment with a power density between 3×10^{16} and 2×10^{17} W/cm^2 [47].

principle and to reduce fluctuations or quantum noise, squeezed states have attracted much attention, especially in quantum optics. Squeezed states in condensed matters including squeezed phonons are also studied and have promising applications such as gravitation-wave detection and quantum communication.

The squeezed phonons were first reported using transient transmission experiment with femtosecond laser pulses for a $KTaO_3$ crystal by Garrett et al. [48]. The measured transmittance coherently oscillated with a frequency close to twice the frequency of the transverse acoustic (2TA) phonon near the zone boundary. These oscillations are due to two phonon-squeezed states excited by the second-order Raman scattering. They estimated that the squeeze factor is approximately 3×10^{-6} of an integrated pulse intensity of $I_0 = 19\ \mu\text{J/cm}^2$.

Generation of phonon-squeezed states by the second-order Raman scattering is also theoretically proposed by Hu and Nori [49]. They considered both the spontaneous Raman and impulsive Raman processes. In the impulsive Raman process, they used a δ function for the optical pulse and the Hamiltonian defined by

$$\hat{H}'_q = \hat{H}_q - \lambda'_q \delta(t) \hat{Q}_q \hat{Q}_{-q}, \tag{5.10}$$

where $\hat{H}_q$ is the Hamiltonian of the phonon system, and λ'_q is the product of the second-order polarizability tensor and electric fields of the optical pulses. By solving the Schrödinger equation with separating the free oscillator terms and the two-phonon creation and annihilation terms, the time-dependent wave function was obtained as

$$\begin{aligned} |\psi_q(t)\rangle &= \exp\left(\frac{\hat{H}_q t}{i\hbar}\right) \exp\left(\frac{i\lambda'_q \hat{H}_q t}{\hbar^2 \omega_q}\right) \\ &\times \exp\left(\zeta'^*_q \hat{b}_q \hat{b}_{-q} - \zeta'_q \hat{b}^\dagger_q \hat{b}^\dagger_{-q}\right) |\psi(0^-)\rangle, \end{aligned} \tag{5.11}$$

where $\zeta'_q = -i\lambda'_q \exp(-i\lambda'_q/\hbar)$. Then, the second-order Raman process causes the two-mode quadrature-squeezed operator on the vacuum state.

In the general squeezed states, the mean value and variance of position oscillate with ω and 2ω, where ω is the phonon frequency, respectively. This feature has also been reported in transient reflection measurements of Bi and GaAs using femtosecond laser pulses with a pump–probe protocol [50]. They measured the amplitude (m) of the oscillatory waveform in transient reflectivity. They repeated the measurement 100 times and calculated the variance, $\sigma^2 = \langle(\langle m\rangle - m)^2\rangle$, at each delay. The transient reflectivity in Bi shows 2.93-THz oscillation, corresponding to the A_{1g} phonon frequency (ω), in amplitude and 2ω frequency in variance. Similar features are observed for the longitudinal optical phonons (frequency of 8.54 THz) in GaAs. They suggested that the observed phase-dependent noise indicates the elliptical shape for the uncertainty contour and the squeezing of phonons.

There are several reports on the squeezed phonons, and further discussions are still continuing [51–53].

5.4 Summary

The coherent phonons are phonons excited by an ultrashort optical pulse and used to study dynamics of optical phonons. A brief history of the experiments and phenomenological theories of the coherent optical phonons is described. The experimental setup and measurement technique for the coherent phonons in diamond are presented as examples. Several selected examples of coherent phonons in semiconductors and semimetals are presented. In addition, the squeezed phonons, which are essential nonclassical states, are described.

References

1. De Silvestri, D., Fujimoto, J.G., Ippen, E.P., Gamble, E.B., Williams, Jr. L.R., Nelson, K.A.: Femtosecond time-resolved measurements of optic phonon dephasing by impulsive stimulated Raman scattering in α-perylene crystal from 20 to 300 K. Chem. Phys. Lett. **116**, 146–152 (1985)
2. Cho, G.C., Kütt, W., Kurz, H.: Subpicosecond time-resolved coherent-phonon oscillation in GaAs. Phys. Rev. Lett. **65**, 764–766 (1990)
3. Cheng, T.K., Brorson, S.D., Kazeroonian, A.S., Moodera, J.S., Dresselhaus, G., Dresselhaus, M.S., Ippen, E.P.: Impulsive excitation of coherent phonons observed in reflection in bismuth and antimony. Appl. Phys. Lett. **57**, 1004–1006 (1990)
4. Chwalek, J.M., Uher, C., Whitaker, J.F., Mourou, G.A., Agostinelli, J.A.: Subpicosecond time-resolved studies of coherent phonon oscillations in thin-film $YBa_2Cu_3O_{6+x}$ (x<0.4). Appl. Phys. Lett. **58**, 980–982 (1991)
5. Misochko, O.V., Hase, M., Ishioka, K., Kitajima, M.: Observation of an amplitude collapse and revival of chirped coherent phonons in bismuth. Phys. Rev. Lett. **92**, 19740 (2004)
6. Katsuki, H., Delagnes, J.D.C., Hosaka, K., Ishioka, K., Chiba, H., Zijlestra, E.S., Garcia, M.E., Takahashi, H., Watanabe, K., Kitajima, M., Matsumoto, Y., Nakamura, K.G., Ohmori, K.: All-optical control and visualization of ultrafast two-dimensional atomic motions in a single crystal of bismuth. Nat. Commun. **4**, 2801 (2013)
7. Cheng, Y.-H., Gao, F.Y., Teitelbaum, S.W., Nelson, K.A.: Coherent control of optical phonons in bismuth. Phys. Rev. B **96**, 134302 (2017)
8. Dekorsy, T., Chi, G.C., Kurz, H.: Coherent phonons in condensed media. Light. Scatt. Solids VIII **76**, 169–209 (2000). and references therein
9. Först, M., Dekorsy, T.: Coherent phonons in bulk and low-dimensional semiconductors. In: De Silvestri, S., Cerullo, G., Lanzani, G. (eds.) Coherent Vibration Dynamics. CRC Press, New York (2008). and references therein
10. Yee, K.J., Lee, K.G., Oh, E., Kim, D.S., Lim, Y.S.: Coherent optical phonons oscillations in bulk GaN excited by far below band gap photons. Phys. Rev. Lett. **88**, 105501 (2002)
11. Hase, M., Kitajima, M., Constantinecu, A.M., Petek, H.: The birth of a quasiparticle in silicon observed in time-frequency space. Nature **426**, 51–54 (2003)
12. Hase, M., Katsuragara, M., Constantinecu, A.M., Petek, H.: Frequency comb generation at terahertz frequencies by coherent phonon excitation in silicon. Nat. Photonics **6**, 243–247 (2012)
13. Yoshino, S., Oohata, G., Mizoguchi, K.: Dynamical Fano-like interference between Rabi oscillations and phonons in a semiconductor microcavity system. Phys. Rev. Lett. **115**, 157402 (2015)
14. Misochko, O.V., Georgiev, N., Dekorsy, T., Helm, M.: Two crossovers in the pseudogap regime of $YBa_2Cu_3O_{7-\delta}$ supreconductors observed by ultrafast spectroscopy. Phys. Rev. Lett. **89**, 067002 (2002)

15. Mansart, B., Boschetto, D., Savoia, A., Rullier-Albenque, F., Forget, A., Colson, D., Rousse, A., Marsi, M.: Observation of a coherent optical phonon in the iron pnictide superconductor $Ba(Fe_{1-x}Co_x)_2As_2$ (x = 0.06 and 0.08). Phys. Rev. B **80**, 172504 (2009)
16. Takahashi, H., Kamihara, Y., Koguchi, H., Atou, T., Hosono, H., Katayama, I., Takeda, J., Kitajima, M., Nakamura, K.G.: Coherent optical phonons in the iron oxypnictide $SmFeAsO_{1-x}F_x$ (x = 0.075). J. Phys. Soc. Jpn. **80**, 013707 (2011)
17. Okano, Y., Katsuki, H., Nakagawa, Y., Takahashi, H., Nakamura, K.G., Ohmori, K.: Optical manipulation of coherent phonons in superconducting $YBa_2Cu_4O_{7-\delta}$ thin films. Faraday Discuss. **153**, 375–382 (2011)
18. Lee, I.H., Yee, K.J., Lee, K.G., Oh, E., Kim, D.S.: Coherent optical phonon mode oscillations in wurtzite ZnO excited by femtosecond pulses. J. Appl. Phys. **93**, 4939–4941 (2003)
19. Hu, J., Misochko, O.V., Takahashi, H., Koguchi, H., Eda, T., Nakamura, K.G.: Ultrafast zone-center coherent lattice dynamics in ferroelectric lithium tantalate. Sci. Technol. Adv. Mater. **12**, 034409 (2011)
20. Wall, S., Wegkamp, D., Foglia, L., Appavoo, K., Nag, J., Haglund, R.F., Jr. Stähler, Wolf, J.: Ultrafast changes in lattice symmetry probed by coherent phonons. Nat. Commun **3**, 721 (2012)
21. Wang, Y., Xu, X., Venkatasubramanian, R.: Reduction in coherent phonon lifetime in Bi_2Te_3/Sb_2Te_3 superlattices. Appl. Phys. Lett. **93**, 113114 (2008)
22. Kamarajyu, N., Kumar, S., Sood, A.K.: Temperature-dependent chirped coherent phonon dynamics in Bi_2Te_3 using high-intensity femtosecond laser pulses. EPL **92**, 47007 (2010)
23. Norimatsu, K., Hu, J., Goto, A., Igarashi, K., Sasagawa, T., Nakamura, K.G.: Coherent optical phonons in a Bi_2Se_3 single crystal measured via transient anisotropic reflectivity. Solid State Commun. **157**, 58–61 (2013)
24. Flock, J., Dekorsy, T., Misochko, O.V.: Coherent lattice dynamics of the topological insulator Bi_2Te_3 probed by ultrafast spectroscopy. Appl. Phys. Lett. **105**, 011902 (2014)
25. Norimatsu, K., Hada, M., Yamamoto, S., Sasagawa, T., Kitajima, M., Kayanuma, Y., Nakamura, K.G.: Dynamics of all Raman active coherent phonons in Sb_2Te_3. J. Appl. Phys. **117**, 143102 (2015)
26. Hu, J., Igarashi, K., Sasagawa, T., Nakamura, K.G., Misochko, O.V.: Femtosecond study of A_{1g} phonons in the strong 3D topological insulators: From pump-probe to coherent control. Appl. Phys. Lett. **112**, 031901 (2018)
27. Mishina, T., Nitta, K., Matsumoto, Y.: Coherent lattice vibration of interlayer shearing mode of graphite. Phys. Rev. B **62**, 2908–2911 (2000)
28. Ishioka, K., Hase, M., Kitajima, M., Petek, H.: Coherent optical phonons in diamond. Appl. Phys. Lett. **89**, 231916 (2006)
29. Gambetta, A., Manzoni, C., Menna, E., Meneghetti, M., Cerullo, G., Lanzani, G., Tretiak, S., Piryatinski, A., Saxena, A., Matin, R.L., Bishop, A.R.: Real-time observation of nonlinear coherent phonon dynamics in singe-walled carbon nanotubes. Nat. Phys. **2**, 515–520 (2006)
30. Kim, J.-H., Nugraha, A.R.T., Booshehri, L.G., Hároz, E.H., Satom, K., Sanders, G.D., Yee, K.-J., Lim, Y.S., Stanton, C.J., Saito, R.: Coherent phonons in carbon nanotubes and graphene. Chem. Phys. **413**, 55–80 (2013)
31. Yan, Y.-X., Gamble Jr. E.B., Nelson, K.A.: Impulsive stimulated scattering: General importance in femtosecond laser pulse interactions with matter, and spectroscopic applications. J. Chem. Phys. **83**, 5391 (1985)
32. Merlin, R.: Generating coherent THz phonons with light pulses. Solid State Commun. **102**, 107–220 (1997)
33. Zeiger, H.J., Vidal, J., Cheng, T.K., Ippen, E.P., Dresselhaus, G., Dresselhaus, M.S.: Theory for displacive excitation of coherent phonons. Phys. Rev. B **45**, 768–778 (1992)
34. Pfeifer, T., Dekorsy, T., Kütt, W., Kurz, H.: Generation mechanism for coherent LO phonons in surface-space-charge fields of III–V-compounds. Appl. Phys. A **55**, 482–488 (1992)
35. Kuznetsov, A.V., Stanton, C.J.: Theory of coherent phonon oscillations in semiconductors. Phys. Rev. Lett. **43**, 3243–3246 (1994)

36. Shinihara, Y., Kawashita, Y., Iwata, J.-I., Yabana, K., Otobe, T., Bertsch, G.F.: First-principles description for coherent phonon generation in diamond. J. Phys.: Condens. Matter **22**, 38412 (2010)
37. Shinihara, Y., Yabana, K., Kawashita, Y., Iwata, J.-I., Otobe, T., Bertsch, G.F.: Coherent phonon generation in time-dependent density functional theory. Phys. Rev. B **82**, 155110 (2010)
38. Sasaki, H., Tanaka, R., Okano, Y., Minami, F., Kayanuma, Y., Shikano, Y., Nakamura, K.G.: Coherent control theory and experiment of optical phonons in diamond. Sci. Rep. **8**, 96909 (2018). https://doi.org/10.1038/s41598-018-27734-1.
39. Hase, M., Mizoguchi, K., Harima, H., Nakashima, S., Tani, M., Sakai, K., Hangyo, M.: Optical control of optical phonons in bismuth filma. Appl. Phys. Lett. **69**, 2474–2476 (1996)
40. Cho, G.C., Dekorsy, T., Bakker, H.J., Hövel, R., Kurz, H.: Generation and relaxation of coherent majority phonons. Phys. Rev. Lett. **77**, 4062–4065 (1996)
41. Hu, J., Zhang, H., Sun, Y., Misochko, O.V., Nakamura, K.G.: Temperature effect on the coupling between coherent longitudinal phonons and plasmons in n-type and p-type GaAs. Phys. Rev. B **97**, 165307 (2018)
42. Sokolowski-Tinten, K., Blome, C., Blums, J., Cavalleri, A., Dietrich, C., Tarasevitch, A., Uschmann, I., Förster, E., Kammler, M., Horn-von-Hoegen, M., von der Linde, D.: Femtosecond X-ray measurement of coherent lattice vibrations near the lindemann stability limit. Nature **422**, 287 (2003)
43. Nakamura, K.G., Ishii, S., Ishitsu, S., Shiokawa, M., Takahashi, H., Dharmalingam, K., Irisawa, J., Hironaka, Y., Ishioka, K., Kitajima, M.: Femtosecond time-resolved X-ray diffraction from optical coherent phonons in CdTe(111) crystal. Appl. Phys. Lett. **93**, 061905 (2008)
44. Johnson, S.L., Beaud, P., Vorobeva, E., Milne, C.J., Murray, É.D., Fahy, S., Ingland, G.: Directly observing squeezed phonon states with femtosecond x-ray diffraction. Phy. Rev. Lett. **102**, 175503 (2009)
45. Harmand, M., Coffee, R., Bionta, M.R., Chollet, M., French, D., Zhu, D., Fritz, D.M., Lemke, H.T., Medvedev, N., Ziaja, B., Toleikis, S., Cammarata, M.: Achieving few-femtosecond time-sorting at har X-ray free-electron lasers. Nat. Photonics **7**, 215–218 (2013)
46. Johnson, S.L., Beaud, P., Möhr-Vorobeva, E., Caviezel, A., Ingold, G., Milne, C.J.: Direct observation of non-fully-symmetric coherent optical phonons by femtosecond x-ray diffraction. Phys. Rev. B **87**, 054301 (2013)
47. Yoshida, M., Fujimoto, Y., Hironaka, Y., Nakamura, K.G., Kondo, K., Ohtani, M., Tsunemi, H.: Generation of picosecond hard x rays by tera watt laser focusing on a copper target. Appl. Phys. Lett. **73**, 2393–2395 (1998)
48. Garrett, G.A., Rojo, A.R., Sood, A.K., Whitaker, J.F., Merlin, R.: Vacuum squeezing of solids: macroscopic quantum states driven by light pulses. Science **275**, 1638–1640 (1997)
49. Hu, X., Nori, F.: Phonon squeezed states generated by second-order Raman scattering. Phys. Rev. Lett. **79**, 4605–4608 (1997)
50. Misochko, O.V., Sakai, K., Nakashima, S.: Phase-dependent noise in femtosecond pump-probe experiments on Bi and GaAs. Phys. Rev. B **61**, 11225–11228 (2000)
51. Misochko, O.V., Hu, J., Nakamura, K.G.: Controlling phonon squeezing correlation via one- and two- phonon interference. Phys. Lett. A **375**, 4141–4146 (2011)
52. Esposito, M., Titimbo, K., Zimmermann, K., Giusti, F., Randi, F., Boschetoo, D., Floreanini, R., Benatti, F., Fausti, D.: Photon number statistics uncover the fluctuations in non-equilibrium lattice dynamics. Nat. Commu. **6**, 10249 (2015)
53. Benatti, F., Esposito, M., Fausi, D., Floreanini, R., Titimbo, K., Zimmermann, K.: Generation and detection of squeezed phonons in lattice dynamics by ultrafast optical excitations. New J. Phys. **19**, 02302 (2017)

Chapter 6
Coherent Phonons: Quantum Theory

In this chapter, we describe a unified quantum mechanical description for the generation and detection of coherent optical phonons using a simple model. Two electronic states coupled with displaced harmonic oscillators, which represent phonons, are used. The dipole interaction between an optical pulse and the system is assumed and treated as a perturbation interaction. Time evolution of the density operator is calculated using the second-order perturbation approximation.

6.1 Generation Mechanism with Displaced Harmonic Oscillator

There are several quantum theories for generating coherent phonons. The different theoretical treatments are often used to describe DECP and ISRS processes [1–5]. Here, we attempt to describe the generation of coherent phonons using a unified quantum mechanical theory [6] in both opaque and transparent conditions.

6.1.1 Hamiltonian

Here, we consider a two-level system for the electronic state and a one-dimensional harmonic oscillator for the optical phonon at the Γ point with a wave vector $k = 0$.[1] Figure 6.1 shows a schematic of the model potential. $|g\rangle$ and $|e\rangle$ indicate the electronic

[1]Quantum mechanical models with the two-level electronic states are often used to study vibrational dynamics of a molecular system [7–10].

K. Nakamura, *Quantum Phononics*, Springer Tracts in Modern Physics 282,
https://doi.org/10.1007/978-3-030-11924-9_6

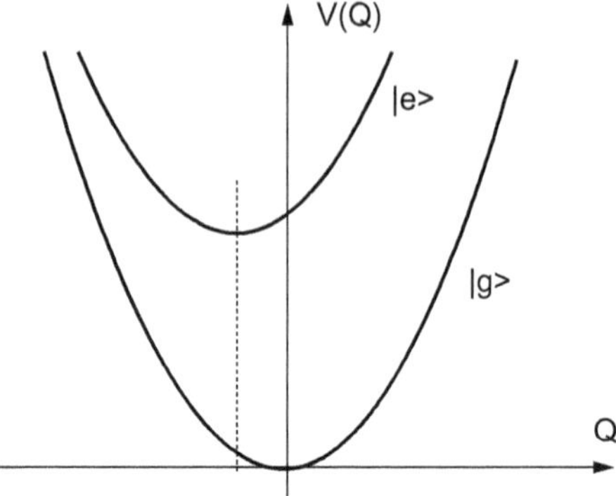

Fig. 6.1 Schematic of the model potential consisting of two electronic states and displaced harmonic potentials

ground and excited states, respectively. The Hamiltonian $\hat{H}_0$ of the materials system is

$$\hat{H}_0 = \hat{H}_g |g\rangle\langle g| + \left(\varepsilon + \hat{H}_e\right) |e\rangle\langle e| \tag{6.1}$$

$$\hat{H}_g = \hbar\omega\hat{b}^\dagger\hat{b} \tag{6.2}$$

$$\hat{H}_e = \hbar\omega\hat{b}^\dagger\hat{b} + \alpha\hbar\omega\left(\hat{b}^\dagger + \hat{b}\right), \tag{6.3}$$

where ω is the angular frequency and $\hat{b}^\dagger$ and $\hat{b}$ are the creation and annihilation operators of the phonon, respectively [6]. $\hat{H}_g$ and $\hat{H}_e$ are Hamiltonians for the harmonic oscillator in the electronic ground and excited states. The zero-point energy of the harmonic oscillator is set to be zero. α represents shift between the harmonic oscillator of the electronic ground and that in the excited states.[2] ε is a energy gap between the electronic ground and excited states. In general, the creation and annihilation operators are dependent on the wave vector k and expressed as $\hat{b}_k^\dagger$ and $\hat{b}_k$. Here, we use only the $\hat{b}^\dagger$ and $\hat{b}$ which represent the operators at $k \sim 0$. The reason is that, according to the phase matching (or momentum conservation), the wave vectors of phonons which can be excited by the optical processes are only those lying close to the Γ-point ($k \sim 0$), because the optical wavelength is much larger than the lattice constant. The eigenmodes of optical phonons form a continuum around $k = 0$. The electromagnetic field of the optical pulse interacts only with atoms within the penetration depth (δL). In opaque conditions, the penetration depth is usually much smaller than the sample thickness. This effect relaxes the condition of phase matching and allows the modes with $\delta k \sim 1/(\delta L)$ around $k = 0$ to be excited. In addition, the spot size of the pump pulse is smaller than the surface area of the crystal, and the phase-matching condition along the lateral direction is also relaxed.

The dipole interaction between the materials system and the incident light is assumed and the interaction Hamiltonian is represented by

[2] α^2 is the Huang–Rhys factor, which is considered to be small ($\alpha^2 < 1$) in a bulk solid.

$$\hat{H}_I(t) = \mu E_0 f(t) \left(e^{-i\Omega t}|e\rangle\langle g| + e^{i\Omega t}|g\rangle\langle e|\right), \tag{6.4}$$

where μ is a transition dipole, E_0 and $f(t)$ are an amplitude and an envelope of electric field of the light pulse. Ω is the angular frequency of the light. The operator $|e\rangle\langle g|$ causes a transition from the electronic ground state $|g\rangle$ to the excited state $|e\rangle$. The light pulse is assumed to be Fourier-transform limited. Here, we use the rotating wave approximation (RWA). The material system Hamiltonian $\hat{H}_0$ is time independent and the interaction Hamiltonian $\hat{H}_I(t)$ is time dependent.

The time evolution of the electron–phonon coupled state is obtained by solving the time-dependent Schrödinger equation as given below:

$$i\hbar\frac{\mathrm{d}}{\mathrm{d}t}|\Psi(t)\rangle = \left(\hat{H}_0 + \hat{H}_I(t)\right)|\Psi(t)\rangle. \tag{6.5}$$

The solution is given formally by using a time-ordered exponential:

$$\begin{aligned}|\Psi(t)\rangle = \exp\left(\frac{1}{i\hbar}\hat{H}_0 t\right)\exp_+\left(\frac{1}{i\hbar}\int_{-\infty}^{t}\hat{H}'_I(t')\mathrm{d}t'\right)\\ \times|\Psi(-\infty)\rangle,\end{aligned} \tag{6.6}$$

with

$$\hat{H}'_I(t') = \exp\left(\frac{i}{\hbar}\hat{H}_0 t\right)\hat{H}_I(t')\exp\left(\frac{-i}{\hbar}\hat{H}_0 t\right), \tag{6.7}$$

where $|\Psi(-\infty)\rangle$ is the Ket vector of the initial state at $t = -\infty$. $\hat{H}'_I(t')$ is given by using (6.1) and (6.3):

$$\begin{aligned}\hat{H}'_I(t) &= \exp\left(\frac{i}{\hbar}\left(\hat{H}_g|g\rangle\langle g| + (\varepsilon + \hat{H}_e\right)|e\rangle\langle e|)t\right)\\ &\quad\times\mu E_0 f(t)(e^{-i\Omega t}|e\rangle\langle g| + e^{i\Omega t}|g\rangle\langle e|)\\ &\quad\times\exp\left(\frac{-i}{\hbar}(\hat{H}_g|g\rangle\langle g| + (\varepsilon + \hat{H}_e)|e\rangle\langle e|)t\right)\\ &= \mu E_0 f(t)\left(\exp\left(\frac{i}{\hbar}(\varepsilon - \hbar\Omega + \hat{H}_e)t\right)|e\rangle\langle g|\exp\left(\frac{-i}{\hbar}\hat{H}_g t\right) + H.c.\right),\end{aligned} \tag{6.8}$$

where we used an expansion of exponential.[3]

[3]When we set $\hat{B} = |e\rangle\langle e|$, the exponential $\exp(\hat{H}_e\hat{B}t)$ is expressed by $\exp(\hat{H}_e\hat{B}t) = \sum_n(\hat{A}t)^n\hat{B}^n$. Since $\hat{B}^2 = (|e\rangle\langle e|)(|e\rangle\langle e|) = |e\rangle\langle e| = \hat{B}$, then $\hat{B}^n = \hat{B}$ for $n \geq 1$. Finally, we get $\exp(\hat{H}_e|e\rangle\langle e|t)|e\rangle\langle g| = \exp(\hat{H}_e t)|e\rangle\langle g|$.

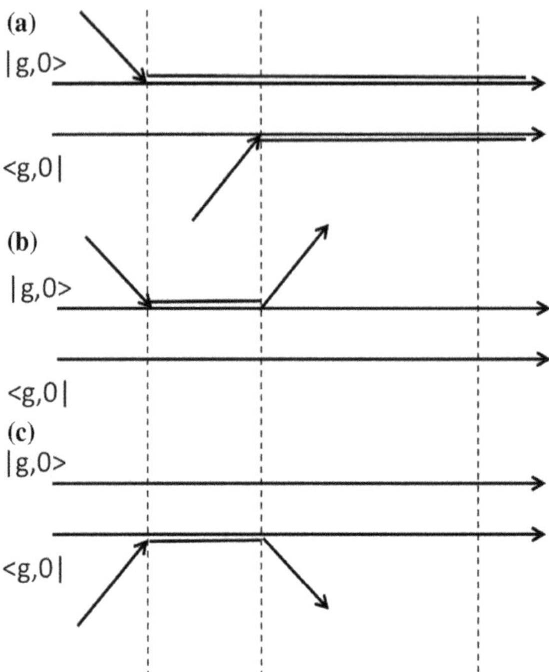

Fig. 6.2 Schematics of Feynman diagrams for the impulsive absorption (**a**) and ISRS (**b** and **c**). The single and double lines represent the electronic ground and excited states, respectively. Time passes from the left side to the right side

6.1.2 Transition Processes

In the abovementioned system, we consider two transition processes: an impulsive absorption process and an impulsive stimulated Raman scattering (ISRS) process. Figure 6.2 shows double-sided Feynman diagrams for the absorption (a) and ISRS (b and c) processes. Time passes from the left side to the right side.

The initial state is set to be the zero-phonon state $|0\rangle$ in the electronic ground state $|g\rangle$, which is expressed by $|g, 0\rangle$. Then, the density matrix of the initial state is $|g, 0\rangle\langle g, 0|$. Here, we consider the second-order perturbation, in which the interaction between the materials system and the light pulse occurs two times. In the absorption process, $|g\rangle$ and $\langle g|$ change to $|e\rangle$ and $\langle e|$ by the first and second interactions, respectively, and the final state becomes $|e\rangle\langle e|$ for the electronic state. The phonon is generated in the electronic excited state. On the other hand, in the ISRS process, $|g\rangle$ changes to $|e\rangle$ by the first interaction and changes to $|g\rangle$ by the second interaction. The final state is in the electronic ground state ($|g\rangle\langle g|$). The phonon is induced in the electronic ground state. It is worth noting that both the absorption and ISRS processes coexist in the quantum mechanics. These processes are expressed with the second-order perturbation calculation.

The IA and ISRS processes refer to the phonon-generation processes via the population and polarization in the electronic excited states, respectively. The DECP mechanism for an opaque condition is included in the IA process. The ISRS process includes not only the ISRS mechanism, which is previously proposed (Chap. 5) for a transparent condition, but also the phonon-generation mechanism in an opaque condition via the photo-induced electronic polarization. The screening of the surface-space-charge field is also included in either the IA or ISRS processes depending on the excitation condition, because both the photoexcited electron–hole pair and the polarization can change the surface-space-charge field, which causes deformation of the potential.[4] Even at the resonance condition, the ISRS process is different from the absorption and emission process, because the population in the electronic state is not induced by the ISRS process.

6.1.3 Second-Order Perturbation

Here, we consider a density operator $\rho(t) \equiv |\psi(t)\rangle\langle\psi(t)|$. The density operator at the initial state is $\rho(0) = |g, 0\rangle\langle g, 0|$. Then, $|\psi(t)\rangle$ at time t is obtained by

$$|\psi(t)\rangle = \exp\left(-\frac{i}{\hbar}\hat{H}_0 t\right)\exp_+\left(\frac{1}{i\hbar}\int_{-\infty}^{t}\hat{H'}_I(t')\mathrm{d}t'\right)|g, 0\rangle. \tag{6.9}$$

We define two operators as

$$\hat{A}^{\dagger}(t) = \exp\left(\frac{i}{\hbar}(\varepsilon - \hbar\Omega + \hat{H}_e)t\right) \tag{6.10}$$

$$\hat{B}(t) = \exp\left(\frac{-i}{\hbar}\hat{H}_g t\right), \tag{6.11}$$

where these operators act on the phonon states. Using these operators, $\hat{H'}_I(t)$ is expressed by

$$\hat{H'}_I(t) = \mu E_0 f(t)\left(\hat{A}^{\dagger}(t)|e\rangle\langle g|\hat{B}(t) + \hat{B}^{\dagger}(t)|g\rangle\langle e|\hat{A}(t)\right). \tag{6.12}$$

Next, we expand the time-ordered exponential in the second-order perturbation and get

[4]The harmonic potential can be deformed under a long-range external field such as the surface-charge field. Supposing a uniform electric field ($F(x) = -\mathrm{d}x$) acting along the phonon coordinate (x), the effective charge field can be treated as uniform in such a short scale. When the external field potential is applied to a harmonic potential $U(x) = kx^2/2$, the potential changes to $U'(x) = U(x) + F(x) = k(x - \mathrm{d}/k)^2/2 - \mathrm{d}^2/(2k)$. Then, the potential minimum position and energy shift are d/k and $-\mathrm{d}^2/(2k)$, respectively. The slope d of the external fields in the electronic excited state is lower than that in the ground state, because the surface screening is suppressed by a electron–hole pair or electronic polarization. Then, the effective harmonic potential in the excited state is displaced from that in the ground state.

$$\begin{aligned}
&\exp_{+}\left(\frac{1}{i\hbar}\int_{-\infty}^{t}\hat{H'}_I(t')\mathrm{d}t'\right)\\
&=1+\frac{1}{i\hbar}\int_{-\infty}^{t}\hat{H'}_I(t')\mathrm{d}t'+\left(\frac{1}{i\hbar}\right)^2\int_{-\infty}^{t}\int_{-\infty}^{t'}\hat{H'}_I(t')\hat{H'}_I(t'')\mathrm{d}t''\mathrm{d}t'\\
&=1+\frac{\mu E_0}{i\hbar}\int_{-\infty}^{t}f(t')\left(\hat{A}^{\dagger}(t')|e\rangle\langle g|\hat{B}(t')+\hat{B}^{\dagger}(t')|g\rangle\langle e|\hat{A}(t')\right)\mathrm{d}t'\\
&+\left(\frac{\mu E_0}{i\hbar}\right)^2\int_{-\infty}^{t}\int_{-\infty}^{t'}f(t')\left(\hat{A}^{\dagger}(t')|e\rangle\langle g|\hat{B}(t')+\hat{B}^{\dagger}(t')|g\rangle\langle e|\hat{A}(t')\right)\\
&\times f(t'')\left(\hat{A}^{\dagger}(t'')|e\rangle\langle g|\hat{B}(t'')+\hat{B}^{\dagger}(t'')|g\rangle\langle e|\hat{A}(t'')\right)\mathrm{d}t''\mathrm{d}t'. \qquad (6.13)
\end{aligned}$$

The third term in the right-hand side is obtained as

$$\begin{aligned}
&\left(\frac{\mu E_0}{i\hbar}\right)^2\int_{-\infty}^{t}\int_{-\infty}^{t''}f(t')f(t'')\big\{\hat{A}^{\dagger}(t')\hat{B}^{\dagger}(t'')|e\rangle\langle e|\hat{B}(t')\hat{A}(t'')\\
&\qquad+\hat{B}^{\dagger}(t')\hat{A}^{\dagger}(t'')|g\rangle\langle g|\hat{A}(t')\hat{B}(t'')\big\}\mathrm{d}t''\mathrm{d}t' \qquad (6.14)
\end{aligned}$$

by using $\langle e|g\rangle = 0$. We define new two operators as

$$\begin{aligned}
\hat{F}(t)&=\int_{-\infty}^{t}f(t')\left(\hat{A}^{\dagger}(t')|e\rangle\langle g|\hat{B}(t')+\hat{B}^{\dagger}(t')|g\rangle\langle e|\hat{A}(t')\right)\mathrm{d}t'\\
\hat{G}(t)&=\int_{-\infty}^{t}\int_{-\infty}^{t'}f(t')f(t'')\big\{\hat{A}^{\dagger}(t')\hat{B}^{\dagger}(t'')|e\rangle\langle e|\hat{B}(t')\hat{A}(t'')\\
&\quad+\hat{B}^{\dagger}(t')\hat{A}^{\dagger}(t'')|g\rangle\langle g|\hat{A}(t')\hat{B}(t'')\big\}\mathrm{d}t''\mathrm{d}t', \qquad (6.15)
\end{aligned}$$

and simplify the ket vector as

$$|\psi(t)\rangle=\exp\left(\frac{-i}{\hbar}\hat{H}_0t\right)\left(1+\frac{\mu E_0}{i\hbar}\hat{F}(t)+\left(\frac{\mu E_0}{i\hbar}\right)^2\hat{G}(t)\right)|g,0\rangle. \qquad (6.16)$$

When the operators ($\hat{F}(t)$ and $\hat{G}(t)$) act on the $|g,0\rangle$ state, we can neglect the terms with $|g\rangle\langle e|$ and $|e\rangle\langle e|$ and the effective operators are

$$\begin{aligned}
\hat{F}(t)&=\int_{-\infty}^{t}f(t')\left(\hat{A}^{\dagger}(t')|e\rangle\langle g|\hat{B}(t')\right)\mathrm{d}t'\\
\hat{G}(t)&=\int_{-\infty}^{t}\int_{-\infty}^{t'}f(t')f(t'')\big\{\hat{B}^{\dagger}(t')\hat{A}^{\dagger}(t'')|g\rangle\langle g|\hat{A}(t')\hat{B}(t'')\big\}\mathrm{d}t''\mathrm{d}t'. \qquad (6.17)
\end{aligned}$$

The density operator is obtained by

$$\begin{aligned}\hat{\rho}(t) &= |\psi(t)\rangle\langle\psi(t)| \\ &= \exp\left(\frac{-i}{\hbar}\hat{H}_0 t\right)\left(1+\left(\frac{\mu E_0}{i\hbar}\right)\hat{F}(t)+\left(\frac{\mu E_0}{i\hbar}\right)^2\hat{G}(t)\right)|g,0\rangle \\ &\times \langle g,0|\left(1+\left(\frac{\mu E_0}{-i\hbar}\right)\hat{F}^\dagger(t)+\left(\frac{\mu E_0}{-i\hbar}\right)^2\hat{G}^\dagger(t)\right)\exp\left(\frac{i}{\hbar}\hat{H}_0 t\right). \quad (6.18)\end{aligned}$$

Here, we focus our attention on the terms with μ^2, which are the second-order interaction with light and denote this as $\hat{\rho}^{(2)}(t)$:

$$\begin{aligned}\hat{\rho}^{(2)}(t) &= \exp\left(\frac{-i}{\hbar}\hat{H}_0 t\right)\left(\frac{\mu E_0}{i\hbar}\right)\hat{F}(t)|g,0\rangle\langle g,0|\left(\frac{\mu E_0}{-i\hbar}\right)\hat{F}^\dagger(t)\exp\left(\frac{i}{\hbar}\hat{H}_0 t\right) \\ &+ \exp\left(\frac{-i}{\hbar}\hat{H}_0 t\right)\left(\frac{\mu E_0}{i\hbar}\right)^2\hat{G}(t)|g,0\rangle\langle g,0|\exp\left(\frac{i}{\hbar}\hat{H}_0 t\right) \\ &+ \exp\left(\frac{-i}{\hbar}\hat{H}_0 t\right)|g,0\rangle\langle g,0|\left(\frac{\mu E_0}{-i\hbar}\right)^2\hat{G}^\dagger(t)\exp\left(\frac{i}{\hbar}\hat{H}_0 t\right), \quad (6.19)\end{aligned}$$

where we denote the terms in the right-hand side as $\hat{\rho}_1^{(2)}(t)$, $\hat{\rho}_2^{(2)}(t)$, and $\hat{\rho}_3^{(2)}(t)$. $\hat{\rho}_1^{(2)}(t)$ corresponds to the impulsive absorption process $\hat{\rho}_2^{(2)}(t)$, and $\hat{\rho}_3^{(2)}(t)$ corresponds to the impulsive stimulated Raman scattering process.

6.1.3.1 Impulsive Absorption

Let us calculate the $\hat{\rho}_1^{(2)}(t)$ term as given below:

$$\begin{aligned}\hat{\rho}_1^{(2)}(t) &= \left(\frac{\mu E_0}{\hbar}\right)^2\exp\left(\frac{-i}{\hbar}\hat{H}_0 t\right)\hat{F}(t)|g,0\rangle\langle g,0|\hat{F}^\dagger(t)\exp\left(\frac{i}{\hbar}\hat{H}_0 t\right) \\ &= \left(\frac{\mu E_0}{\hbar}\right)^2\exp\left(\frac{-i}{\hbar}\hat{H}_0 t\right)\int_{-\infty}^{t}\mathrm{d}t' f(t')\left(\hat{A}^\dagger(t')|e\rangle\langle g|\hat{B}(t')\right)|g,0\rangle \\ &\times \langle g,0|\int_{-\infty}^{t}\mathrm{d}t' f(t')\left(\hat{B}^\dagger(t')|e\rangle\langle g|\hat{A}(t')\right)\exp\left(\frac{i}{\hbar}\hat{H}_0 t\right). \quad (6.20)\end{aligned}$$

The former term in the right-hand side becomes

$$\begin{aligned}&\exp\left(\frac{-i}{\hbar}\hat{H}_0 t\right)\int_{-\infty}^{t}\mathrm{d}t' f(t')\left(\hat{A}^\dagger(t')|e\rangle\langle g|\hat{B}(t')\right)|g,0\rangle \\ &= \int_{-\infty}^{t}\mathrm{d}t' f(t')\exp\left(\frac{-i}{\hbar}\left(\hat{H}_g|g\rangle\langle g|+\left(\varepsilon+\hat{H}_e|e\rangle\langle e|\right)\right)t\right)\end{aligned}$$

$$
\begin{aligned}
&\times \exp\left(\frac{i}{\hbar}\left(\varepsilon - \hbar\Omega + \hat{H}_e\right)t'\right)\exp\left(\frac{-i}{\hbar}\hat{H}_g t'\right)|e,0\rangle \\
&= \int_{-\infty}^{t} dt' f(t') \exp\left(\frac{-i}{\hbar}\left(\varepsilon - \hbar\Omega + \hat{H}_e\right)(t-t')\right)\exp\left(\frac{-i}{\hbar}\hbar\Omega t\right)|e,0\rangle,
\end{aligned} \tag{6.21}
$$

where we used

$$
\exp\left(\frac{-i}{\hbar}\hat{H}_g t'\right)|0\rangle = \left(1 + \left(\hbar\omega\hat{b}^\dagger\hat{b}\right) + \frac{1}{2}\left(\hbar\omega\hat{b}^\dagger\hat{b}\right)^2 + \cdots\right)|0\rangle = |0\rangle, \tag{6.22}
$$

because $\hat{b}^\dagger\hat{b}|0\rangle = 0$. A similar calculation gives the latter half in the right-hand side in (6.20) as

$$
\langle e,0| \int_{-\infty}^{t} dt' f(t') \exp\left(\frac{i}{\hbar}\left(\varepsilon - \hbar\Omega + \hat{H}_e\right)(t-t')\right)\exp\left(\frac{i}{\hbar}\hbar\Omega t\right). \tag{6.23}
$$

Then, we get the density operator as

$$
\begin{aligned}
\hat{\rho}_1^{(2)}(t) = \left(\frac{\mu E_0}{\hbar}\right)^2 &\int_{-\infty}^{t} dt' f(t') \exp\left(\frac{-i}{\hbar}\left(\varepsilon - \hbar\Omega + \hat{H}_e\right)(t-t')\right)|e,0\rangle \\
&\langle e,0| \int_{-\infty}^{t} dt'' f(t'') \exp\left(\frac{i}{\hbar}\left(\varepsilon - \hbar\Omega + \hat{H}_e\right)(t-t'')\right),
\end{aligned} \tag{6.24}
$$

which shows the system is in the electronic excited state at time t. The expectation value for the electronic excitation state is

$$
\begin{aligned}
\langle e|\hat{\rho}_1^{(2)}(t)|e\rangle &= \left(\frac{\mu E_0}{\hbar}\right)^2 \int_{-\infty}^{t} dt' f(t') \exp\left(\frac{-i}{\hbar}\left(\varepsilon - \hbar\Omega + \hat{H}_e\right)(t-t')\right)|0\rangle \\
&\quad \langle 0| \int_{-\infty}^{t} dt'' f(t'') \exp\left(\frac{i}{\hbar}\left(\varepsilon - \hbar\Omega + \hat{H}_e\right)(t-t'')\right) \\
&= \left(\frac{\mu E_0}{\hbar}\right)^2 |\psi_p(t)\rangle\langle\psi_p(t)|,
\end{aligned} \tag{6.25}
$$

where $|\psi_p(t)\rangle$ is the phonon ket vector:

$$
|\psi_p(t)\rangle = \int_{-\infty}^{t} dt' f(t') \exp\left(\frac{-i}{\hbar}\left(\varepsilon - \hbar\Omega + \hat{H}_e\right)(t-t')\right)|0\rangle. \tag{6.26}
$$

The final electronic state is the excited state and this process corresponds to the absorption process.

6.1.3.2 Impulsive Stimulated Raman Scattering Process

Next, we calculate the $\hat{\rho}_2^{(2)}(t)$ term in a similar way as

$$\begin{aligned}\hat{\rho}_2^{(2)}(t) &= \left(\frac{\mu E_0}{\hbar}\right)^2 \exp\left(\frac{-i}{\hbar}\hat{H}_0 t\right)\hat{G}(t)|g,0\rangle\langle g,0|\exp\left(\frac{i}{\hbar}\hat{H}_0 t\right)\\ &= \left(\frac{\mu E_0}{\hbar}\right)^2 \exp\left(\frac{-i}{\hbar}\hat{H}_0 t\right)\int_{-\infty}^{t}\int_{-\infty}^{t'} f(t')f(t'')\big\{\hat{B}^\dagger(t')\hat{A}^\dagger(t'')\\ &\times |g\rangle\langle g|\hat{A}(t')\hat{B}(t'')\big\}\mathrm{d}t''\mathrm{d}t'|g,0\rangle\langle g,0|\exp\left(\frac{i}{\hbar}\hat{H}_0 t\right).\end{aligned} \tag{6.27}$$

Here

$$\hat{B}^\dagger(t')\hat{A}^\dagger(t'')|g\rangle = \exp\left(\frac{i}{\hbar}\hat{H}_g t'\right)\exp\left(\frac{i}{\hbar}\left(\varepsilon - \hbar\Omega + \hat{H}_e\right)t''\right)|g\rangle, \tag{6.28}$$

and

$$\langle g|\hat{A}(t')\hat{B}(t'') = \langle g|\exp\left(\frac{-i}{\hbar}\left(\varepsilon - \hbar\Omega + \hat{H}_e\right)t'\right)\exp\left(\frac{-i}{\hbar}\hat{H}_g t''\right), \tag{6.29}$$

then, we get

$$\begin{aligned}\hat{B}^\dagger(t')\hat{A}^\dagger(t'')|g\rangle\langle g|\hat{A}(t')\hat{B}(t'') &= \exp\left(\frac{i}{\hbar}\hat{H}_g t'\right)\\ \times\exp\left(\frac{-i}{\hbar}\left(\varepsilon - \hbar\Omega + \hat{H}_e\right)(t'-t'')\right)&\exp\left(\frac{-i}{\hbar}\hat{H}_g t''\right)|g\rangle\langle g|.\end{aligned} \tag{6.30}$$

The integral is then expressed by

$$\begin{aligned}&\int_{-\infty}^{t}\int_{-\infty}^{t'} f(t')f(t'')\exp\left(\frac{i}{\hbar}\hat{H}_g t'\right)\exp\left(\frac{-i}{\hbar}\left(\varepsilon - \hbar\Omega + \hat{H}_e\right)(t'-t'')\right)\\ &\qquad\times\exp\left(\frac{-i}{\hbar}\hat{H}_g t''\right)\mathrm{d}t''\mathrm{d}t'|g,0\rangle\\ &= \int_{-\infty}^{t}\int_{-\infty}^{t'} f(t')f(t'')\exp\left(\frac{i}{\hbar}\hat{H}_g t'\right)\\ &\qquad\times\exp\left(\frac{-i}{\hbar}\left(\varepsilon - \hbar\Omega + \hat{H}_e\right)(t'-t'')\right)\mathrm{d}t'\mathrm{d}t''|g,0\rangle\end{aligned} \tag{6.31}$$

In addition,

$$\langle g,0|\exp\left(\frac{i}{\hbar}\hat{H}_0 t\right) = |g,0\rangle \exp\left(\frac{i}{\hbar}\hat{H}_g|g\rangle\langle g|t\right) = \langle g,0|, \tag{6.32}$$

then, we get

$$\begin{aligned}\hat{\rho}_2^{(2)}(t) = &\left(\frac{\mu E_0}{i\hbar}\right)^2 \exp\left(\frac{-i}{\hbar}\hat{H}_0 t\right)\int_{-\infty}^{t}\int_{-\infty}^{t'} f(t')f(t'')\exp\left(\frac{i}{\hbar}\hat{H}_g t'\right)\\ &\exp\left(\frac{-i}{\hbar}\left(\varepsilon - \hbar\Omega + \hat{H}_e\right)(t'-t'')\right)\mathrm{d}t'\mathrm{d}t''|g,0\rangle\langle g,0|\\ = &\left(\frac{\mu E_0}{i\hbar}\right)^2 \int_{-\infty}^{t}\int_{-\infty}^{t'} f(t')f(t'')\exp\left(\frac{-i}{\hbar}\hat{H}_g(t-t')\right)\\ &\exp\left(\frac{-i}{\hbar}\left(\varepsilon - \hbar\Omega + \hat{H}_e\right)(t'-t'')\right)\mathrm{d}t'\mathrm{d}t''|g,0\rangle\langle g,0|. \end{aligned}\tag{6.33}$$

Therefore, the expectation value for the electronic states, which corresponds to the trace for the electronic states, is

$$\begin{aligned}\langle g|\hat{\rho}_2^{(2)}(t)|g\rangle = &\langle g|\left(\frac{\mu E_0}{i\hbar}\right)^2 \exp\left(\frac{-i}{\hbar}\hat{H}_0 t\right)\int_{-\infty}^{t}\int_{-\infty}^{t'} f(t')f(t'')\exp\left(\frac{i}{\hbar}\hat{H}_g t'\right)\\ &\exp\left(\frac{-i}{\hbar}\left(\varepsilon - \hbar\Omega + \hat{H}_e\right)(t'-t'')\right)\mathrm{d}t'\mathrm{d}t''|g,0\rangle\langle g,0||g\rangle\\ = &\left(\frac{\mu E_0}{i\hbar}\right)^2 \int_{-\infty}^{t}\int_{-\infty}^{t'} f(t')f(t'')\exp\left(\frac{-i}{\hbar}\hat{H}_g(t-t')\right)\\ &\exp\left(\frac{-i}{\hbar}\left(\varepsilon - \hbar\Omega + \hat{H}_e\right)(t'-t'')\right)\mathrm{d}t'\mathrm{d}t''|0\rangle\langle 0|\\ = &|\psi_p(t)\rangle\langle 0|, \end{aligned}\tag{6.34}$$

where $|\psi_p(t)\rangle$ is the ket for a phonon state

$$\begin{aligned}|\psi_p(t)\rangle = &\int_{-\infty}^{t}\int_{-\infty}^{t'} f(t')f(t'')\exp\left(\frac{-i}{\hbar}\hat{H}_g(t-t')\right)\\ &\exp\left(\frac{-i}{\hbar}\left(\varepsilon - \hbar\Omega + \hat{H}_e\right)(t'-t'')\right)\mathrm{d}t'\mathrm{d}t''|0\rangle. \end{aligned}\tag{6.35}$$

$\hat{\rho}_3^{(2)}(t)$ is also calculated in a similar way to give

$$\hat{\rho}_3^{(2)}(t) = |0\rangle\langle\psi_p(t)|, \tag{6.36}$$

which is the Hermitian conjugate of $\hat{\rho}_2^{(2)}(t)$. In the paths represented by $\hat{\rho}_2^{(2)}(t)$ and $\hat{\rho}_3^{(2)}(t)$, the initial and final states are in the electronic ground state and the light interaction occurs for the bra or ket state. These paths correspond to the ISRS process.

6.1.4 *Time Evolution of Phonon Amplitude*

The expectation value of the phonon amplitude is calculated by $\langle Q(t)\rangle$ as given below:

$$\begin{aligned}\langle Q(t)\rangle &= Tr(\hat{Q}\hat{\rho}^{(2)}(t)) \\ &= Tr(\hat{Q}\hat{\rho}_1^{(2)}(t)) + Tr(\hat{Q}\hat{\rho}_2^{(2)}(t)) + Tr(\hat{Q}\hat{\rho}_3^{(2)}(t)).\end{aligned} \tag{6.37}$$

6.1.4.1 Impulsive Absorption Process

The phonon ket in the electronic excited state which is expressed by (6.26) as

$$|\psi_P(t)\rangle = \int_{-\infty}^{t} f(t')\exp\left(\frac{-i}{\hbar}(\varepsilon - \hbar\Omega)(t-t')\right)\exp\left(\frac{-i}{\hbar}\hat{H}_e(t-t')\right)dt'|0\rangle. \tag{6.38}$$

The Hamiltonian for the harmonic oscillator in the $|e\rangle$ is

$$\begin{aligned}\hat{H}_e &= \hbar\omega\hat{b}^\dagger\hat{b} + \alpha\hbar\omega(\hat{b}^\dagger + \hat{b}) \\ &= \hat{D}^\dagger(\alpha)\hat{H}_g\hat{D}(\alpha) - \hbar\omega\alpha^2,\end{aligned} \tag{6.39}$$

where $\hat{D}(\alpha)$ is the displacement operator and expressed by $\hat{D}(\alpha) \equiv \exp\left(\alpha\hat{b}^\dagger - \alpha^*\hat{b}\right)$. This relation is shown by

$$\begin{aligned}\hat{D}^\dagger(\alpha)\hat{H}_g\hat{D}(\alpha) &= \hbar\omega\left(\hat{D}^\dagger(\alpha)\hat{b}^\dagger\hat{b}\hat{D}(\alpha)\right) \\ &= \hbar\omega\left(\hat{D}^\dagger(\alpha)\hat{b}^\dagger\hat{D}\hat{D}^\dagger\hat{b}\hat{D}(\alpha)\right) \\ &= \hbar\omega\left(\hat{b}^\dagger + \alpha\right)\left(\hat{b} + \alpha\right) \\ &= \hat{H}_e + \hbar\omega\alpha^2.\end{aligned} \tag{6.40}$$

Using the displacement operator, the coherent state $|\alpha\rangle$ is expressed by $|\alpha\rangle = \hat{D}(\alpha)|0\rangle$. Using these relations, we get

$$\exp\left(\frac{-i}{\hbar}\hat{H}_e t\right) = \hat{D}^{\dagger}(\alpha)\exp\left(\frac{-i}{\hbar}\hat{H}_g t\right)\hat{D}(\alpha)\exp\left(i\omega\alpha^2 t\right)$$
$$= \exp\left(i\omega\alpha^2 t\right)\hat{D}^{\dagger}(\alpha)\exp\left(-i\omega\hat{b}^{\dagger}\hat{b}\right)\hat{D}(\alpha), \qquad (6.41)$$

and $|\psi_p(t)\rangle$ is obtained as

$$|\psi_p(t)\rangle = \int_{-\infty}^{t} f(t')\exp\left(-\frac{i}{\hbar}(\varepsilon - \hbar\Omega)(t-t')\right)\exp\left(i\omega\alpha^2(t-t')\right)$$
$$\times \hat{D}^{\dagger}(\alpha)\exp\left(-i\omega\hat{b}^{\dagger}\hat{b}(t-t')\right)\hat{D}(\alpha)\mathrm{d}t'|0\rangle$$
$$= \int_{-\infty}^{t} f(t')\exp\left(-\frac{i}{\hbar}(\varepsilon - \hbar\Omega - \alpha^2\hbar\omega)(t-t')\right)$$
$$\times \hat{D}^{\dagger}(\alpha)|\alpha e^{-i\omega(t-t')}\rangle\mathrm{d}t', \qquad (6.42)$$

where $|\alpha e^{-i\omega(t-t')}\rangle$ is the time-dependent coherent state. By setting a new parameter $B \equiv -(\varepsilon - \hbar\Omega - \alpha^2\hbar\omega)/(\hbar\omega)$, the expectation value of the phonon amplitude in the electronic excited state is obtained as

$$\bar{Q} = \langle\psi_p(t)|\sqrt{\frac{\hbar}{2\omega}}\left(\hat{b}^{\dagger} + \hat{b}\right)|\psi_p(t)\rangle$$
$$= \sqrt{\frac{\hbar}{2\omega}}\int_{-\infty}^{t}\int_{-\infty}^{t} f(t'')f(t')e^{-i\omega B(t-t'')}e^{i\omega B(t-t')}$$
$$\times \langle\alpha e^{-i\omega(t-t'')}|\hat{D}(\alpha)\left(\hat{b}^{\dagger} + \hat{b}\right)\hat{D}^{\dagger}(\alpha)|\alpha e^{-i\omega(t-t')}\rangle\mathrm{d}t'\mathrm{d}t''$$
$$= \sqrt{\frac{\hbar}{2\omega}}\int_{-\infty}^{t}\int_{-\infty}^{t} f(t'')f(t')e^{-i\omega B(t'-t'')}$$
$$\times \langle\alpha e^{-i\omega(t-t'')}|\left(\hat{b}^{\dagger} - \alpha + \hat{b} - \alpha\right)|\alpha e^{-i\omega(t-t')}\rangle\mathrm{d}t'\mathrm{d}t''$$
$$= \sqrt{\frac{\hbar}{2\omega}}\int_{-\infty}^{t}\int_{-\infty}^{t} f(t'')f(t')e^{-i\omega B(t'-t'')}$$
$$\times \alpha\left(e^{i\omega(t-t'')} + e^{-i\omega(t-t')} - 2\right)\langle\alpha e^{-i\omega(t-t'')}|\alpha e^{-i\omega(t-t')}\rangle\mathrm{d}t'\mathrm{d}t''. \qquad (6.43)$$

The time-dependent coherent state is expressed by

$$|\alpha e^{-i\omega(t-t')}\rangle = e^{-|\alpha|^2/2}\sum_{n=0}^{\infty}\frac{\alpha^n}{\sqrt{n!}}e^{-i\omega(t-t')n}|n\rangle, \qquad (6.44)$$

and

$$\langle \alpha e^{-i\omega(t-t'')}| = e^{-|\alpha|^2/2} \sum_{m=0}^{\infty} \langle m| \frac{\alpha^m}{\sqrt{m!}} e^{i\omega(t-t'')m}. \tag{6.45}$$

Using these relations, we get

$$\begin{aligned}\langle \alpha e^{-i\omega(t-t'')}|\alpha e^{-i\omega(t-t')}\rangle &= e^{-|\alpha|^2} \sum_{n=0}^{\infty} \frac{\alpha^{2n}}{n!} e^{-i\omega(t''-t')n} \\ &= \exp\left(-|\alpha|^2\right) \exp\left(|\alpha|^2 e^{-i\omega(t''-t')}\right) \\ &= \exp\left(|\alpha|^2 \left(e^{i\omega(t'-t'')} - 1\right)\right). \end{aligned} \tag{6.46}$$

Then, $\bar{Q}$ is obtained as

$$\begin{aligned}\bar{Q} = &\sqrt{\frac{\hbar}{2\omega}} \int_{-\infty}^{t} \int_{-\infty}^{t} f(t') f(t'') e^{-i\omega B(t'-t'')} \times \alpha \left(e^{i\omega(t-t'')} + e^{-i\omega(t-t')} - 2\right) \\ &\times \exp\left(|\alpha|^2 \left(e^{i\omega(t'-t'')} - 1\right)\right) \mathrm{d}t' \mathrm{d}t''. \end{aligned} \tag{6.47}$$

Finally, we obtain the expectation value of the phonon amplitude by inserting the form of B as

$$\begin{aligned}\langle Q_A(t)\rangle &= Tr[\hat{Q}\hat{\rho}_2^{(2)}] = \left(\frac{\mu E_0}{\hbar}\right)^2 \bar{Q} \\ &= \alpha \left(\frac{\mu E_0}{\hbar}\right)^2 \sqrt{\frac{\hbar}{2\omega}} \int_{-\infty}^{t} \int_{-\infty}^{t} f(t') f(t'') \left(e^{i\omega(t-t'')} + e^{-i\omega(t-t')} - 2\right) \\ &\times \exp\left(-\alpha^2 \left(1 + i\omega(t'-t'') - e^{i\omega(t'-t'')}\right)\right) e^{i(\varepsilon - \hbar\Omega)(t'-t'')/\hbar} \mathrm{d}t' \mathrm{d}t''. \end{aligned} \tag{6.48}$$

If the displacement α between the excited-state and ground-state potentials is small ($\alpha \ll 1$), we regard $\alpha^2 \approx 0$ and get a simple form as

$$\begin{aligned}\langle Q_A(t)\rangle = &\alpha \left(\frac{\mu E_0}{\hbar}\right)^2 \sqrt{\frac{\hbar}{2\omega}} \int_{-\infty}^{t} \int_{-\infty}^{t} f(t') f(t'') \left(e^{i\omega(t-t'')} + e^{-i\omega(t-t')} - 2\right) \\ &\times e^{i(\varepsilon - \hbar\Omega)(t'-t'')/\hbar} \mathrm{d}t' \mathrm{d}t''. \end{aligned} \tag{6.49}$$

6.1.4.2 Impulsive Stimulated Raman Scattering Process

In the ISRS process, the final state is the electronic ground state, which is given by

$$\langle g|\hat{\rho}_2^{(2)}(t) + \hat{\rho}_3^{(2)}(t)|g\rangle = -\left(\frac{\mu E_0}{\hbar}\right)^2 \left(|\psi_p(t)\rangle\langle 0| + |0\rangle\langle\psi_p(t)|\right). \quad (6.50)$$

The $|\psi_p(t)\rangle$ is expressed using the initial state $|0\rangle$ at $t = -\infty$ as

$$|\psi_p(t)\rangle = \int_{-\infty}^{t}\int_{-\infty}^{t'} f(t'')f(t')\exp\left(\frac{-i}{\hbar}\hat{H}_g(t-t')\right) \times \exp\left(\frac{-i}{\hbar}(\varepsilon - \hbar\Omega + \hat{H}_e)(t'-t'')\right)\exp\left(\frac{-i}{\hbar}\hat{H}_g t'\right)\mathrm{d}t'\mathrm{d}t''|0\rangle. \quad (6.51)$$

The expectation value of the phonon amplitude induced by the ISRS is

$$\langle Q_R(t)\rangle = -\left(\frac{\mu E_0}{\hbar}\right)^2 \sqrt{\frac{\hbar}{2\omega}}\left(\langle 0|\hat{b}|\psi_p(t)\rangle + \langle\psi_p(t)|\hat{b}^\dagger|0\rangle\right). \quad (6.52)$$

The first term of this equation is calculated by

$$\langle 0|\hat{b}|\psi_p(t)\rangle = \int_{-\infty}^{t}\int_{-\infty}^{t'} f(t')f(t'')\langle 0|\hat{b}\exp\left(\frac{-i}{\hbar}\hat{H}_g(t-t')\right) \times \exp\left(\frac{-i}{\hbar}(\varepsilon - \hbar\Omega)(t'-t'')\right)\exp\left(\frac{-i}{\hbar}\hat{H}_e(t'-t'')\right)\mathrm{d}t''\mathrm{d}t'|0\rangle. \quad (6.53)$$

Using the relationship

$$\exp\left(\frac{-i}{\hbar}\hat{H}_e(t'-t'')\right) = e^{i\omega\alpha^2(t'-t'')}\hat{D}^\dagger(\alpha)e^{-i\omega\hat{b}^\dagger\hat{b}(t'-t'')}\hat{D}(\alpha), \quad (6.54)$$

we get

$$\exp\left(\frac{-i}{\hbar}\hat{H}_e(t'-t'')\right)|0\rangle = e^{i\omega\alpha^2(t'-t'')}\hat{D}^\dagger(\alpha)|\alpha e^{-i\omega(t'-t'')}\rangle. \quad (6.55)$$

On the other hand, there is a relationship

$$e^{i\omega\hat{b}^\dagger\hat{b}t}\hat{b}e^{-i\omega\hat{b}^\dagger\hat{b}t} = e^{-i\omega t}\hat{b} \quad (6.56)$$

and then

$$\hat{b}e^{-i\omega\hat{b}^\dagger\hat{b}t} = e^{-i\omega t}e^{-i\omega\hat{b}^\dagger\hat{b}t}\hat{b}. \quad (6.57)$$

Using this relation, we get

$$\hat{b}\exp\left(\frac{-i}{\hbar}\hat{H}_g(t-t')\right) = e^{-i\omega(t-t')}\exp\left(\frac{-i}{\hbar}\hat{H}_g(t-t')\right)\hat{b} \quad (6.58)$$

and

$$\langle 0|\hat{b}\exp\left(\frac{-i}{\hbar}\hat{H}_g(t-t')\right)\exp\left(\frac{-i}{\hbar}\hat{H}_e(t'-t'')\right)|0\rangle$$
$$= e^{-i\omega(t-t')}e^{-i\omega\alpha^2(t'-t''))}\langle 0|\hat{b}\hat{D}^\dagger(\alpha)|\alpha e^{-i\omega(t'-t'')}\rangle. \tag{6.59}$$

This braket is simplified as

$$\begin{aligned}\langle 0|\hat{b}\hat{D}^\dagger(\alpha)|\alpha e^{-i\omega(t'-t'')}\rangle &= \langle 0|\hat{D}^\dagger(\alpha)\hat{D}(\alpha)\hat{b}\hat{D}^\dagger(\alpha)|\alpha e^{-i\omega(t'-t'')}\rangle \\ &= \langle 0|\hat{D}^\dagger(\alpha)(\hat{b}-\alpha)|\alpha e^{-i\omega(t'-t'')}\rangle \\ &= \langle\alpha|(\hat{b}-\alpha)|\alpha e^{-i\omega(t'-t'')}\rangle \\ &= \alpha\left(e^{-i\omega(t'-t'')}-1\right)\langle\alpha|\alpha e^{-i\omega(t'-t'')}\rangle. \end{aligned} \tag{6.60}$$

Using these equations, we get

$$\begin{aligned}\langle 0|\hat{b}|\psi_p(t)\rangle &= \int_{-\infty}^{t}\mathrm{d}t'\int_{-\infty}^{t'}\mathrm{d}t''f(t')f(t'')e^{-i\omega(t-t')}e^{-i\omega\alpha^2(t'-t'')} \\ &\times\alpha\left(e^{-i\omega(t'-t'')}-1\right)\exp\left(\alpha^2(e^{-i\omega(t'-t'')}-1)\right)e^{-i(\varepsilon-\hbar\Omega)(t'-t'')/\hbar} \\ &= \alpha\int_{-\infty}^{t}\mathrm{d}t'\int_{-\infty}^{t'}\mathrm{d}t''f(t')f(t'')\left(e^{-i\omega(t-t'')}-e^{-i\omega(t-t')}\right) \\ &\times\exp\left(-\alpha^2\left(1-i\omega(t'-t'')+e^{-i\omega(t'-t'')}\right)\right)e^{-i(\varepsilon-\hbar\Omega)(t'-t'')/\hbar}. \end{aligned} \tag{6.61}$$

Finally, we get the expectation value of the phonon amplitude as

$$\begin{aligned}\langle Q_R(t)\rangle &= \alpha\left(\frac{\mu E_0}{\hbar}\right)^2\sqrt{\frac{\hbar}{2\omega}}\int_{-\infty}^{t}\mathrm{d}t'\int_{-\infty}^{t'}\mathrm{d}t''f(t')f(t'')\left(e^{-i\omega(t-t'')}-e^{-i\omega(t-t')}\right) \\ &\times\exp\left(-\alpha^2\left(1-i\omega(t'-t'')+e^{-i\omega(t'-t'')}\right)\right)e^{-i(\varepsilon-\hbar\Omega)(t'-t'')/\hbar}+c.c. \end{aligned} \tag{6.62}$$

For a small α case, the equation is simplified as

$$\begin{aligned}\langle Q_R(t)\rangle = \alpha\left(\frac{\mu E_0}{\hbar}\right)^2\sqrt{\frac{\hbar}{2\omega}}\int_{-\infty}^{t}\mathrm{d}t'\int_{-\infty}^{t'}\mathrm{d}t''f(t')f(t'') \\ \left(e^{-i\omega(t-t'')}-e^{-i\omega(t-t')}\right)e^{-i(\varepsilon-\hbar\Omega)(t'-t'')/\hbar}+c.c. \end{aligned} \tag{6.63}$$

6.1.5 Gaussian Pulse Case

The time evolution of the phonon amplitude is obtained by numerical calculation of (6.48) and (6.63). For some simple cases, for example, the Gaussian function of the envelope of an optical pulse, we obtain analytical solutions. Here, we use an optical pulse $E(t)$ with a Gaussian envelope function:

$$E(t) = E_0 f(t) \left(\exp(i\Omega t) + \exp(-i\Omega t)\right) \tag{6.64}$$

$$f(t) = \frac{1}{\sqrt{\pi}\sigma} \exp\left(\frac{-t^2}{\sigma^2}\right), \tag{6.65}$$

where E_0 is the peak intensity of the electric field, Ω is the optical frequency, σ is the pulse width, and $1/(\sqrt{\pi}\sigma)$ is a normalization factor. We hereafter set $(\mu E_0/\hbar)^2 = 1$ and $\alpha \ll 1$ for simplifying calculations.

6.1.5.1 Resonance Condition

At the resonance condition of the excitation optical pulse ($\varepsilon - \hbar\Omega = 0$), the amplitude of the phonons excited via the absorption process is expressed

$$\begin{aligned}\langle Q_A(t)\rangle = \alpha\sqrt{\frac{\hbar}{2\omega}} \int_{-\infty}^{t} \mathrm{d}t' \int_{-\infty}^{t} \mathrm{d}t'' f(t'') f(t') \\ \times \left(e^{-i\omega(t-t')} + e^{i\omega(t-t'')} - 2\right).\end{aligned} \tag{6.66}$$

Here, $f(t)$ is a real function and $e^{-i\omega(t-t')} = \cos\omega(t-t') - i\sin\omega(t-t')$ and $e^{i\omega(t-t'')} = \cos\omega(t-t'') - i\sin\omega(t-t'')$. Because t' and t'' are independent and have the same integration range, the sine-function terms become zero. Then, we get

$$\begin{aligned}\langle Q_A(t)\rangle = \alpha\sqrt{\frac{\hbar}{2\omega}} \int_{-\infty}^{t} \mathrm{d}t' \int_{-\infty}^{t} \mathrm{d}t'' f(t'') f(t') \\ \times \left(\cos\omega(t-t') + \cos\omega(t-t'') - 2\right).\end{aligned} \tag{6.67}$$

The similar calculation gives the expected amplitude of the phonons excited by the ISRS process which is given by

$$\begin{aligned}\langle Q_R(t)\rangle &= \alpha\sqrt{\frac{\hbar}{2\omega}} \int_{-\infty}^{t} \mathrm{d}t' \int_{-\infty}^{t'} \mathrm{d}t'' f(t') f(t'') \\ &\quad \times \left(e^{-i\omega(t-t')} - e^{-i\omega(t-t'')}\right) + c.c. \\ &= 2\alpha\sqrt{\frac{\hbar}{2\omega}} \int_{-\infty}^{t} \mathrm{d}t' \int_{-\infty}^{t'} \mathrm{d}t'' f(t') f(t'') \\ &\quad \times \left(\cos\omega(t-t') - \cos\omega(t-t'')\right).\end{aligned} \tag{6.68}$$

Impulsive absorption process:
For a long time delay condition ($t \gg \sigma$), the upper limit of the integral can be replaced by ∞. We integrate separately for t' and t'', because they are independent:

$$\int_{-\infty}^{\infty} \mathrm{d}t' f(t') e^{-i\omega(t-t')} = \frac{1}{\sqrt{\pi}\sigma} e^{-i\omega t} \int_{-\infty}^{\infty} e^{-t'^2/\sigma^2} e^{i\omega t'} \mathrm{d}t'. \tag{6.69}$$

By using the Gaussian integral, we can calculate as

$$\begin{aligned} \int_{-\infty}^{\infty} e^{-t'^2/\sigma^2} e^{i\omega t'} \mathrm{d}t' &= \int_{-\infty}^{\infty} \exp\left(-\frac{(t' - i\sigma\omega/2)^2}{\sigma^2}\right) \mathrm{d}t' \exp\left(\frac{-\sigma^2\omega^2}{4}\right) \\ &= \sqrt{\pi}\sigma \exp\left(\frac{-\sigma^2\omega^2}{4}\right). \end{aligned} \tag{6.70}$$

Then, the phonon amplitude is obtained as

$$\begin{aligned} \langle Q_A(t) \rangle &= \alpha\sqrt{\frac{\hbar}{2\omega}} \left(\exp\left(\frac{-\sigma^2\omega^2}{4}\right) e^{-i\omega t} + \exp\left(\frac{-\sigma^2\omega^2}{4}\right) e^{i\omega t} - 2 \right) \\ &= 2\alpha\sqrt{\frac{\hbar}{2\omega}} \left(\exp\left(\frac{-\sigma^2\omega^2}{4}\right) \cos(\omega t) - 1 \right) \\ &= Q_0 \left(\exp\left(\frac{-\sigma^2\omega^2}{4}\right) \cos(\omega t) - 1 \right), \end{aligned} \tag{6.71}$$

where $Q_0 \equiv 2\alpha\sqrt{\hbar/2\omega}$. The result means that oscillation of the phonon amplitude is cosine-like. If we take a short pulse limit ($\sigma\omega \to 0$), the phonon amplitude is expressed by

$$\langle Q_A(t) \rangle = Q_0 \left(\cos\omega t - 1 \right). \tag{6.72}$$

On the other hand, the phonon amplitude decreases along with increasing pulse width.

ISRS process:
We also consider the $t \gg \sigma$ condition and set the upper limit of the integration ∞:

$$\langle Q_R(t) \rangle = Q_0 e^{-i\omega t} \int_{-\infty}^{\infty} \mathrm{d}t' \int_{-\infty}^{\infty} \mathrm{d}t'' f(t') f(t'') \left(e^{i\omega t'} - e^{i\omega t''} \right) + c.c. \tag{6.73}$$

Here, we set new parameters (s and u) by

$$s \equiv \left(t' + t''\right)/2 \tag{6.74}$$

$$u \equiv t' - t'', \tag{6.75}$$

which have the integral range $[-\infty, \infty]$ for s and $[0, \infty]$ for u, respectively. t' and t'' are expressed by $t' = s + u/2$ and $t'' = s - u/2$, respectively. Using these parameters, we get

$$f(t')f(t'')\left(e^{i\omega t'} - e^{i\omega t''}\right) = \frac{1}{\pi\sigma^2}\exp\left(\frac{-(t'^2 + t''^2)}{\sigma^2}\right)\left(e^{i\omega t'} - e^{i\omega t''}\right), \tag{6.76}$$

and

$$t'^2 + t''^2 = (s + u/2)^2 + (s - u/2)^2 = 2s^2 + u^2/2 \tag{6.77}$$

$$\begin{aligned} e^{i\omega t'} - e^{i\omega t''} &= e^{i\omega(s+u/2)} - e^{i\omega(s-u/2)} \\ &= e^{i\omega s}2i\sin(\omega u/2). \end{aligned} \tag{6.78}$$

Using these equations, we get

$$\begin{aligned} \langle Q_R(t)\rangle = Q_0\frac{e^{-i\omega t}}{\pi\sigma^2}\int_{-\infty}^{\infty}\mathrm{d}s\int_0^{\infty}\mathrm{d}u \\ \times\exp\left(-\frac{2s^2}{\sigma^2} - \frac{u^2}{2\sigma^2}\right)e^{i\omega s}2i\sin\left(\frac{\omega u}{2}\right) + c.c. \end{aligned} \tag{6.79}$$

Here, we use

$$\int_{-\infty}^{\infty}\mathrm{d}s\exp\left(-\frac{2s^2}{\sigma^2} + i\omega s\right) = \frac{\sqrt{\pi}}{\sqrt{2}\sigma}\exp\left(\frac{-\sigma^2\omega^2}{8}\right) \tag{6.80}$$

and

$$\int_0^{\infty}\mathrm{d}u\exp\left(-\frac{u^2}{2\sigma^2}\right)\sin\left(\frac{\omega u}{2}\right) = \sqrt{2}\sigma\exp\left(\frac{-\sigma^2\omega^2}{8}\right)\int_0^{\sigma\omega 2\sqrt{2}} t^2\mathrm{d}t. \tag{6.81}$$

This integral is defined by the pulse width and the phonon frequency and constant for a given condition. When we set this constant value as

$$A \equiv Q_0\frac{4}{\sqrt{\pi}}\exp\left(\frac{-\sigma^2\omega^2}{4}\right)\int_0^{\sigma\omega 2\sqrt{2}} t^2\mathrm{d}t \tag{6.82}$$

the phonon amplitude is given by

$$\langle Q_R(t)\rangle = A\sin\omega t. \tag{6.83}$$

The amplitude of phonons induced by ISRS oscillates sine-like, though that of the impulsive absorption oscillates cosine-like. It is worth noting that the optical phonon amplitude goes to zero for the short pulse limit ($\sigma\omega \to 0$), because the integral A goes to zero.

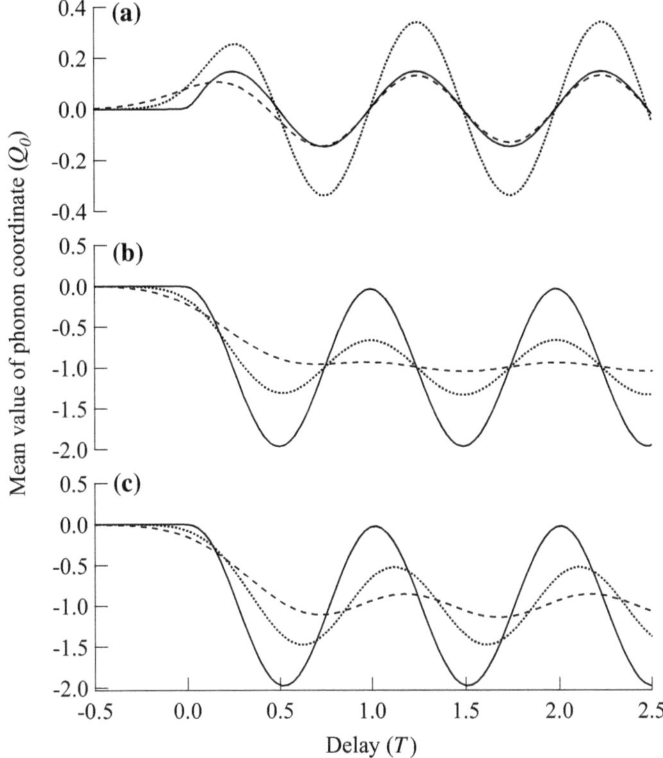

Fig. 6.3 Time evolution of the mean value of the phonon amplitude with pulse width (FWHM) of $0.1T$ (solid curves), $0.55T$ (dotted curves), and $0.9T$ (dashed curves) for the vibrational period T of the optical phonon. The bottom axis represents the pump–probe delay in unit of T. **a** represents $\langle Q_R \rangle$ excited by ISRS process, **b** represents $\langle Q_A \rangle$ excited by IA process. The total amplitude $\langle Q \rangle = \langle Q_R \rangle + \langle Q \rangle_A$ is shown in **c**. This figure is obtained by modifying Fig. 3 of the paper, K.G. Nakamura et al., Physical Review B 92, 144304 (2015)

6.1.5.2 Numerical Calculation: Pulse Width Dependence

Figure 6.3 shows the numerical results of the time evolution of the mean value of the phonon amplitude and its pulse width dependence. We examined three pulse width conditions ($0.1T$, $0.55T$, and $0.9T$), where T is the vibrational period of the optical phonon. For $\langle Q_A(t) \rangle$, the approximate formula (6.71) agrees well with the numerical results in the region of time after the passage of the pulse (delay larger than T). Violent oscillations of the coherent phonons change to a gradual adaption of a new equilibrium as the pulse width becomes large. The phonon amplitude decreases as the pulse width increases. On the other hand, for $\langle Q_R(t) \rangle$, the phonon amplitude at the pulse width of $0.55T$ is larger than that at $0.1T$ and $0.9T$.

In order to discriminate experimentally between $\langle Q_A(t) \rangle$ and $\langle Q_R(t) \rangle$, it is necessary to use the electronic-state-selective measurement of the coherent phonons.

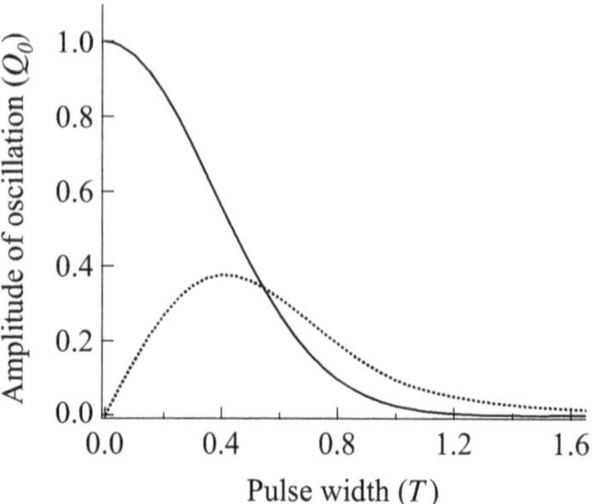

Fig. 6.4 The dependence of the pulse width on the phonon amplitudes $\langle Q_A(t)\rangle$ (solid curve) and $\langle Q_R(t)\rangle$ (dotted curve). This figure is obtained by modifying Fig. 4 of the paper, K.G. Nakamura et al., Physical Review B 92, 144304 (2015)

However, in the commonly used transient reflectivity or transmissivity measurements, the electronic state is not identified. We get information only on the total value $\langle Q(t)\rangle = \langle Q_A(t)\rangle + \langle Q_R(t)\rangle$, which is shown in Fig. 6.3c. $\langle Q(t)\rangle$ shows an oscillation with the frequency ω, which is almost cosine-like at short pulse width such as $0.1T$. As the pulse width increases, the contribution from the ISRS process increases, and the initial phase of the oscillation changes.

In Fig. 6.4, the phonon amplitudes are plotted for IA (solid curve) and ISRS (dotted curve) as a function of the pulse width (FWHM). The phonon amplitude by ISRS process increases with an increase of the pulse width, and reaches maximum at $0.55T$ and decreases at a larger pulse width. The dominant process changes from IA to ISRS around a pulse width of $0.55T$.

6.1.5.3 Nonresonant Condition: Detuning Effect

We consider detuning effects on the coherent phonon generation, when a frequency of the optical pulse does not match the energy gap between electronic two levels. We restrict small α value ($\alpha \ll 1$) and neglect the perturbation terms higher than the second order. The detuning is defined by $\xi \equiv (\varepsilon - \hbar\Omega)/\hbar$.

For the impulsive absorption process, the phonon amplitude is obtained analytically as a similar way for the resonant case:

$$\langle Q_A(t)\rangle = Q_0 \exp\left(\frac{-\sigma^2\xi^2}{2}\right)\left(\exp\left(\frac{-\sigma^2(\omega^2+2\xi\omega)}{4}\right)\cos(\omega t) - 1\right). \quad (6.84)$$

This shows that the phonon amplitude decreases as the detuning increases.

For the ISRS process, the calculation is a little bit complicated. The phonon amplitude is expressed by

$$\langle Q_R(t)\rangle = \frac{1}{\sqrt{2\pi}} e^{-\sigma^2\omega^2/8} e^{-i\omega t} \int_0^\infty \mathrm{d}u e^{-u^2/2\sigma^2} \\ \times \left(e^{-i\omega u/2} - e^{-i\omega u/2}\right) e^{-i\xi u} + c.c. \tag{6.85}$$

The integral term gives a complex number, then an initial phase of the phonon-amplitude oscillation is dependent on the detuning. Here, we evaluate the large detuning case ($\xi\sigma \gg 1$) compared to the pulse width. In this case, the real part of the integral term is negligibly small. For the positive value of η, the integral is calculated by

$$\int_0^\infty \mathrm{d}u \exp\left(-\frac{u^2}{2\sigma^2} - i\eta u\right) = \sqrt{\frac{\pi}{2}}\sigma \exp\left(-\frac{\sigma^2\eta^2}{2}\right) \\ - i\sqrt{2}\sigma \exp\left(-\frac{\sigma^2\eta^2}{2}\right) \int_0^{\sigma\eta/\sqrt{2}} e^{t^2} \mathrm{d}t. \tag{6.86}$$

The real part is smaller than the imaginary part value for the $\eta\sigma \gg 1$ case. Then, we can get

$$\langle Q_R(t)\rangle = B \sin(\omega t), \tag{6.87}$$

where

$$B \equiv Q_0\sqrt{2} \exp\left(\frac{-\sigma^2\omega^2}{8}\right)\left(D\left(\xi + \frac{\omega}{2}\right) - D\left(\xi - \frac{\omega}{2}\right)\right). \tag{6.88}$$

$D(x)$ is the Dawson function and expressed by

$$D(x) \equiv \exp\left(\frac{-\sigma^2 x^2}{2}\right) \int_0^{\sigma x/\sqrt{2}} e^{t^2} \mathrm{d}t. \tag{6.89}$$

For the large x case, $D(x)$ is expanded with x^{-1} as $D(x) \sim a_0 + a_1 x^{-1} + a_2 x^{-2} + \cdots$, and $\mathrm{d}D(x)/\mathrm{d}x$ is expressed by

$$\frac{\mathrm{d}}{\mathrm{d}x} D(x) = -\sigma^2 x D(x) + \frac{\sigma}{\sqrt{2}}. \tag{6.90}$$

These equations give

$$D(x) \simeq \frac{1}{\sqrt{2}\sigma x}. \tag{6.91}$$

Finally, we get the approximate form of B:

$$B \simeq -Q_0\sqrt{\frac{2}{\pi}}\exp\left(\frac{-\sigma^2\omega^2}{8}\right)\frac{\omega}{\sigma\xi^2}. \tag{6.92}$$

6.1.5.4 Numerical Calculation: Detuning Dependence

Figure 6.5 shows the numerical results of the time evolution of phonon amplitude $\langle Q_A(t)\rangle$, $\langle Q_R(t)\rangle$, and $\langle Q(t)\rangle$ with a pulse width of $0.1T$ with detuning of $\Delta E -$ $0, 3\hbar\omega, 5\hbar\omega$, and $10\hbar\omega$. The phonon amplitude $\langle Q_A(t)\rangle$ approaches to zero as the detuning increases. $\langle Q_R(t)\rangle$ starts to move in the same direction as $\langle Q_A(t)\rangle$ at large detuning.

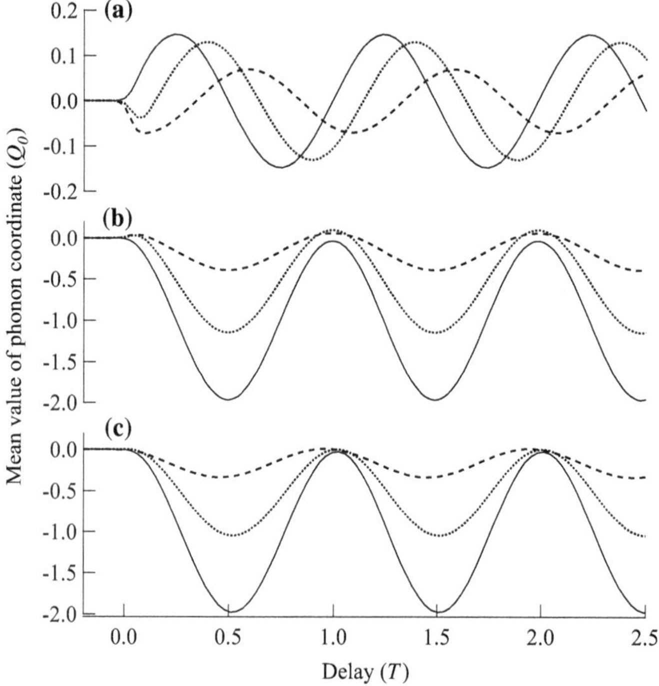

Fig. 6.5 Time evolution of the phonon amplitude with detuning ΔE of $0\hbar\omega$ (solid curves: no detuning), $3\hbar\omega$ (dotted curves), and $5\hbar\omega$ (dashed curves). **a** represents $\langle Q_R(t)\rangle$, **b** represents $\langle Q_A(t)\rangle$, and **c** represents $\langle Q(t)\rangle$. T is the vibrational period of the phonon. The pulse width was set to $0.1T$. This figure is obtained by modifying Fig. 5 of the paper, K.G. Nakamura et al., Physical Review B 92, 144304 (2015)

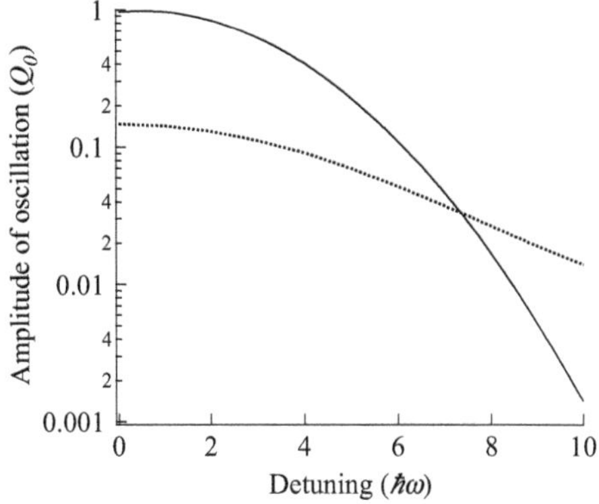

Fig. 6.6 The detuning dependence of the phonon amplitude for $\langle Q_A(t)\rangle$ (solid curve) and $\langle Q_R(t)\rangle$ (dotted curve). The pulse width was $0.1T$. This figure is obtained by modifying Fig. 6 of the paper, K.G. Nakamura et al., Physical Review B 92, 144304 (2015)

Figure 6.6 shows the phonon amplitude as a function of the detuning. Both the phonon amplitudes $\langle Q_R(t)\rangle$ and $\langle Q_A(t)\rangle$ decrease as the detuning increases. For large detuning, the phonon amplitude $\langle Q_R(t)\rangle$ becomes larger than $\langle Q_A(t)\rangle$ since no light absorption occurs, and the dominant generation process of the coherent phonons is subject to the ISRS process.

6.2 Four-Level Model and Double-Sided Feynman Diagrams

In a weak coupling ($\alpha < 1$) case, the generation of coherent phonons can be calculated using a four-level system consisting of two electronic states ($|g\rangle$ and $|e\rangle$) and zero-phonon and one-phonon states ($|0\rangle$ and $|1\rangle$). There are four levels: $|g\rangle|0\rangle$, $|g\rangle|1\rangle$,$|e\rangle|0\rangle$, and $|e\rangle|1\rangle$. The four-level model is easily used to calculate dynamics compared to that of the displaced harmonic oscillators (Sect. 6.1), while the both show the same results in a weak coupling case.

6.2.1 Impulsive Absorption Process

Transition processes are represented using double-sided Feynman diagrams. Figure 6.7 shows an impulsive absorption process with the $|g, 0\rangle \to |e, 1\rangle$ and $\langle g, 0| \to \langle e, 0|$ transitions. The final state ($|e, 1\rangle\langle e, 0|$) is in the electronic excited state with vibrational polarization. We set the initial state $|g, 0\rangle\langle g, 0|$ at $t = -\infty$. At time t'', $|g, 0\rangle$ transitions to $|e, 1\rangle$ via the dipole interaction with the incident pulse and

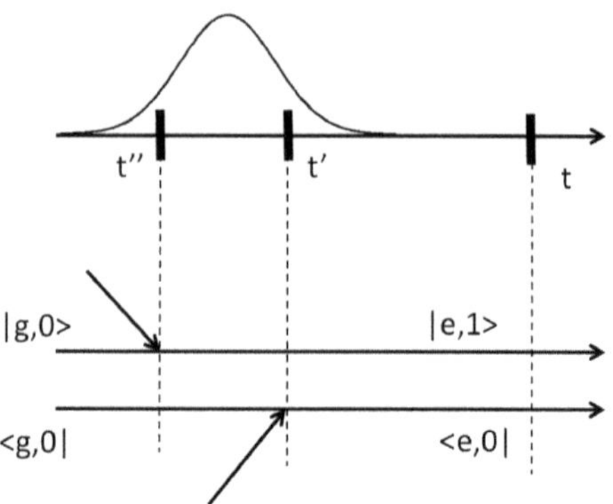

Fig. 6.7 Diagram of the impulsive absorption process. Time flows from the left to the right. The upper part of the figure shows an envelope of the optical pulse. Interaction between the optical pulse and the system occurs at time t″ and t′

the electron–phonon coupling ($\alpha\hat{b}^{\dagger}$). The total interaction causes the multiplicative coefficient

$$\alpha\left(\frac{\mu E(t'')}{i\hbar}\right)=\alpha\left(\frac{\mu E_0}{i\hbar}\right)f(t'')e^{-i\Omega t''}. \tag{6.93}$$

At time between t'' and t', the state $|e,1\rangle\langle g,0|$ has a time evolution factor of

$$e^{-i(\varepsilon/\hbar+\omega)(t'-t'')}. \tag{6.94}$$

At time t', $\langle g,0|$ transfers to $\langle e,0|$ via the dipole interaction with the incident pulse. This interaction causes the multiplicative coefficient

$$\left(\frac{\mu E^*(t')}{-i\hbar}\right)=\left(\frac{\mu E_0}{-i\hbar}\right)f(t')e^{i\Omega t'}. \tag{6.95}$$

After time t', the state $|e,1\rangle\langle e,0|$ propagates with a factor of

$$e^{-i\omega(t-t')}. \tag{6.96}$$

Through these processes, we get the time evolution of the density matrix by

$$\begin{aligned}\rho_1(t:t',t'')&=\alpha\left(\frac{\mu E_0}{i\hbar}\right)e^{-i\Omega t''}f(t'')e^{-i(\varepsilon/\hbar+\omega)(t'-t'')}\\&\quad\times\left(\frac{\mu E_0}{-i\hbar}\right)f(t')e^{i\Omega t'}e^{-i\omega(t-t')}|e,1\rangle\langle e,0|\\&=\alpha\left(\frac{\mu E_0}{\hbar}\right)^2 f(t'')f(t')e^{-i\Delta(t'-t'')}e^{-i\omega(t-t'')}|e,1\rangle\langle e,0|,\end{aligned} \tag{6.97}$$

where $\Delta = \varepsilon/\hbar - \Omega$. By integrating $\rho(t : t', t'')$ over t'' and t', we get

$$\rho_1(t) = \alpha \left(\frac{\mu E_0}{\hbar}\right)^2 \int_{-\infty}^{t} \mathrm{d}t' \int_{-\infty}^{t} \mathrm{d}t'' f(t'') f(t') e^{-i\Delta(t'-t'')} e^{-i\omega(t-t'')} \times |e, 1\rangle\langle e, 0|. \tag{6.98}$$

Another path is $\langle g, 0| \to \langle e, 0|$ at time t'' and $|g, 0\rangle \to |e, 1\rangle$ at time t'. Its density matrix is given by

$$\rho_2(t) = \alpha \left(\frac{\mu E_0}{\hbar}\right)^2 \int_{-\infty}^{t} \mathrm{d}t' \int_{-\infty}^{t} \mathrm{d}t'' f(t'') f(t') e^{i\Delta(t'-t'')} e^{-i\omega(t-t')} \times |e, 1\rangle\langle e, 0|. \tag{6.99}$$

In addition to these two paths, their Hermitian conjugate paths are possible. The total density matrix $\rho(t)$ is obtained by $\rho(t) = \rho_1(t) + \rho_2(t) + H.c.$

At the resonance condition ($\Delta = 0$), the density matrix is expressed as

$$\begin{aligned}\rho(t) = \alpha \left(\frac{\mu E_0}{\hbar}\right)^2 &\int_{-\infty}^{t} \mathrm{d}t' \int_{-\infty}^{t} \mathrm{d}t'' f(t'') f(t') \\ &\times \left(\left(e^{-i\omega(t-t'')} + e^{-i\omega(t-t')}\right) |e, 1\rangle\langle e, 0| \right. \\ &\left. + \left(e^{i\omega(t-t'')} + e^{i\omega(t-t')}\right) |e, 0\rangle\langle e, 1| \right). \end{aligned} \tag{6.100}$$

In the absorption processes, phonons are excited in the electronic excited state, in which the potential energy is expressed by the displaced harmonic oscillator, and the annihilation and creation operators are expressed by $\hat{b} - \alpha$ and $\hat{b}^\dagger - \alpha$, respectively. The phonon coordinate operator $\hat{Q}_A$ is expressed by

$$\hat{Q}_A = \sqrt{\frac{\hbar}{2\omega}} \left(\hat{b} + \hat{b}^\dagger - 2\alpha\right). \tag{6.101}$$

Therefore, the $|e, 0\rangle\langle e, 0|$ state also contributes to the shifted position, and the density operator $\rho'(t)$ for the $|g, 0\rangle\langle g, 0| \to |e, 0\rangle\langle e, 0|$ transition is

$$\begin{aligned}\rho'(t) &= \left(\frac{\mu E_0}{\hbar}\right)^2 \int_{-\infty}^{t} \mathrm{d}t' \int_{-\infty}^{t} \mathrm{d}t'' f(t'') f(t') \left(e^{-i\Delta(t'-t'')} + e^{-i\Delta(t''-t')}\right) |e, 0\rangle\langle e, 0| \\ &= \left(\frac{\mu E_0}{\hbar}\right)^2 \int_{-\infty}^{t} \mathrm{d}t' \int_{-\infty}^{t} \mathrm{d}t'' f(t'') f(t') |e, 0\rangle\langle e, 0|. \end{aligned} \tag{6.102}$$

The expectation value of the phonon amplitude $\langle \hat{Q}_A \rangle$ is calculated as

$$\begin{aligned}\langle \hat{Q}_A(t) \rangle &= Tr((\rho(t) + \rho'(t))\hat{Q}_A) \\ &= \alpha \left(\frac{\mu E_0}{\hbar} \right)^2 \int_{-\infty}^{t} dt' \int_{-\infty}^{t} dt'' f(t'') f(t') \\ &\quad \times \left(\cos(\omega(t - t')) + \cos(\omega(t - t'')) - 2 \right). \end{aligned} \tag{6.103}$$

This is consistent with the result (6.49) with $\varepsilon = \hbar\omega$.

6.2.2 *Impulsive Stimulated Raman Scattering Process*

Next, we consider the ISRS process for exciting the coherent phonons in the electronic ground state.

Figure 6.8a and b show the double-sided Feynman diagrams of the impulsive stimulated Raman scattering process with the $|g, 0\rangle \to |e, 1\rangle \to |g, 1\rangle$ and $|g, 0\rangle \to |e, 0\rangle \to |g, 1\rangle$ transitions, respectively. The final state ($|g, 1\rangle\langle g, 0|$) is in the electronic ground state with vibrational polarization. We set the initial state $|g, 0\rangle\langle g, 0|$ at $t = -\infty$. In the path shown by Fig. 6.8a, at time t'', $|g, 0\rangle$ changes to $|e, 1\rangle$ via the dipole interaction with the incident pulse and the electron–phonon coupling ($\alpha\hat{b}^\dagger$). The total interaction causes the multiplicative coefficient

$$\alpha \left(\frac{\mu E(t'')}{i\hbar} \right) = \alpha \left(\frac{\mu E_0}{i\hbar} \right) f(t'') e^{-i\Omega t''}, \tag{6.104}$$

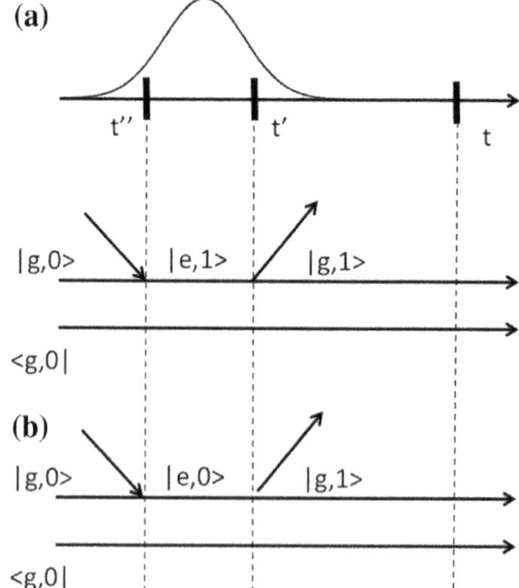

Fig. 6.8 Diagram of the impulsive stimulated Raman scattering process. **a** passes through $|e, 1\rangle$ state and **b** passes through the $|e, 0\rangle$ state. Time flows from the left to the right. The upper part of the figure shows an envelope of the optical pulse. Interaction between the optical pulse and the system occurs at time t″ and t′

which is the same as that in the IA process (Fig. 6.7). At time between t'' and t', the state $|e, 1\rangle\langle g, 0|$ has a time evolution factor of

$$e^{-i(\varepsilon/\hbar+\omega)(t'-t'')}. \tag{6.105}$$

At time t', $|e, 1\rangle$ changes to $|g, 1\rangle$ via the dipole interaction with the incident pulse. This interaction causes the multiplicative coefficient

$$\left(\frac{\mu E^*(t')}{i\hbar}\right) = \left(\frac{\mu E_0}{i\hbar}\right) f(t')e^{i\Omega t'}, \tag{6.106}$$

because the light goes out from the material system. After time t', the state $|g, 1\rangle\langle g, 0|$ evolves with a factor of

$$e^{-i\omega(t-t')}. \tag{6.107}$$

Through these processes, we get the time evolution of the density matrix as

$$\begin{aligned} \rho_1(t : t', t'') &= \alpha \left(\frac{\mu E_0}{i\hbar}\right) e^{-i\Omega t''} f(t'')e^{-i(\varepsilon/\hbar+\omega)(t'-t'')} \\ &\times \left(\frac{\mu E_0}{i\hbar}\right) f(t')e^{i\Omega t'} e^{-i\omega(t-t')} |g, 1\rangle\langle g, 0| \\ &= -\alpha \left(\frac{\mu E_0}{\hbar}\right)^2 f(t'')f(t')e^{-i\Delta(t'-t'')}e^{-i\omega(t-t'')}|g, 1\rangle\langle g, 0|. \end{aligned} \tag{6.108}$$

By integrating $\rho(t : t', t'')$ over t'' and t', we get

$$\begin{aligned} \rho_1(t) = -\alpha \left(\frac{\mu E_0}{\hbar}\right)^2 \int_{-\infty}^{t} \mathrm{d}t' \int_{-\infty}^{t'} \mathrm{d}t'' f(t'')f(t')e^{-i\Delta(t'-t'')}e^{-i\omega(t-t'')} \\ \times |g, 1\rangle\langle g, 0|. \end{aligned} \tag{6.109}$$

Another path (Fig. 6.8b) is $|g, 0\rangle \to |e, 0\rangle$ at time t'' and $|e, 0\rangle \to |g, 1\rangle$ at time t'. At time t'', $|g, 0\rangle$ changes to $|e, 0\rangle$ via the dipole interaction with the incident pulse. The interaction causes the multiply coefficient

$$\left(\frac{\mu E(t'')}{i\hbar}\right) = \left(\frac{\mu E_0}{i\hbar}\right) f(t'')e^{-i\Omega t''}. \tag{6.110}$$

At time between t'' and t', the state $|e, 0\rangle\langle g, 0|$ has a time evolution factor of

$$e^{-i(\varepsilon/\hbar)(t'-t'')}. \tag{6.111}$$

At time t', $|e, 0\rangle$ changes to $|g, 1\rangle$ via the dipole interaction with the incident pulse and the electron–phonon coupling. This interaction causes the multiplicative coefficient

$$-\alpha\left(\frac{\mu E^*(t')}{i\hbar}\right) = -\alpha\left(\frac{\mu E_0}{i\hbar}\right) f(t')e^{i\Omega t'}. \tag{6.112}$$

It is worth noting that the electron–phonon coupling interaction is $-\alpha\hat{b}$ for the transition from the electronic excited to the ground state, because the ground-state harmonic potential shifts in the $-\alpha$ direction. After time t', the state $|g, 1\rangle\langle g, 0|$ evolves with a factor of

$$e^{-i\omega(t-t')}. \tag{6.113}$$

Through these processes, we get the time evolution of the density matrix as

$$\begin{aligned}\rho_2(t:t',t'') &= \left(\frac{\mu E_0}{i\hbar}\right) e^{-i\Omega t''} f(t'')e^{-i(\varepsilon/\hbar)(t'-t'')}(-\alpha)\\ &\quad\times \left(\frac{\mu E_0}{i\hbar}\right) f(t')e^{i\Omega t'}e^{-i\omega(t-t')}|g, 1\rangle\langle g, 0|\\ &= \alpha\left(\frac{\mu E_0}{\hbar}\right)^2 f(t'')f(t')e^{-i\Delta(t'-t'')}e^{-i\omega(t-t')}|g, 1\rangle\langle g, 0|.\end{aligned} \tag{6.114}$$

By integrating $\rho_2(t:t',t'')$ over t'' and t', we get

$$\begin{aligned}\rho_2(t) = \alpha\left(\frac{\mu E_0}{\hbar}\right)^2 \int_{-\infty}^{t} \mathrm{d}t' \int_{-\infty}^{t'} \mathrm{d}t'' f(t'')f(t')e^{-i\Delta(t'-t'')}e^{-i\omega(t-t')}\\ \times |g, 1\rangle\langle g, 0|.\end{aligned} \tag{6.115}$$

In addition to these two paths, their Hermitian conjugate paths are possible. The total density matrix $\rho(t)$ is obtained by $\rho(t) = \rho_1(t) + \rho_2(t) + H.c.$ At the resonance condition ($\Delta = 0$), the density matrix $\rho(t)$ is obtained to be

$$\begin{aligned}\rho(t) = \alpha\left(\frac{\mu E_0}{\hbar}\right)^2 \int_{-\infty}^{t} \mathrm{d}t' \int_{-\infty}^{t'} \mathrm{d}t'' f(t'')f(t')\\ \times \left(e^{-i\omega(t-t')} - e^{-i\omega(t-t'')}\right)|g, 1\rangle\langle g, 0| + H.c.\end{aligned} \tag{6.116}$$

In the stimulated Raman scattering process, phonons are excited in the electronic ground state, and the phonon coordinate operator $\hat{Q}_R$ is expressed by

$$\hat{Q}_R = \sqrt{\frac{\hbar}{2\omega}}\left(\hat{b} + \hat{b}^\dagger\right). \tag{6.117}$$

The expectation value of the phonon amplitude $\langle \hat{Q}_R \rangle$ is calculated as

$$\begin{aligned}\langle \hat{Q}_R(t) \rangle =& Tr(\rho(t)\hat{Q}_R) \\ =& 2\alpha\sqrt{\frac{\hbar}{2\omega}}\left(\frac{\mu E_0}{\hbar}\right)^2 \int_{-\infty}^{t} \mathrm{d}t' \int_{-\infty}^{t} \mathrm{d}t'' f(t'') f(t') \left(\cos(\omega(t-t'')) - \cos(\omega(t-t'))\right).\end{aligned} \tag{6.118}$$

This is the same as (6.63).

6.3 Extension to Band Model

The two-level system for the electronic state is extended to a two band model, in which electronic excited states with a different wave vector k have different energy levels [11]. The one-dimensional harmonic potentials are also assumed for the optical phonons. The Hamiltonian is expressed as

$$\begin{aligned}\hat{H}_0 &= \hat{H}_g|g\rangle\langle g| + \sum_k \hat{H}_k|k\rangle\langle k| \\ \hat{H}_g &= \hbar\omega\hat{b}^\dagger\hat{b} \\ \hat{H}_k &= \varepsilon_k + \hbar\omega\hat{b}^\dagger\hat{b} + \alpha\hbar\omega(\hat{b}+\hat{b}^\dagger),\end{aligned} \tag{6.119}$$

where $|k\rangle$ is the state for which an electron with wave vector k is excited from $|g\rangle$ to the conduction band with the excitation energy of ε_k. The creation and annihilation operators of the LO phonon at the Γ-point with energy $\hbar\omega$ are denoted $\hat{b}^\dagger$ and $\hat{b}$, respectively. The deformation potential interaction having dimensionless coupling constant α, which is independent of the k vector, is assumed. Using the rotating wave approximation, the interaction Hamiltonian between pump pulse and the electronic state is given by

$$\hat{H}_I(t) = \sum_k \mu_k E_0 f(t)\left(e^{-i\Omega t}|k\rangle\langle g| + e^{i\Omega t}|g\rangle\langle k|\right), \tag{6.120}$$

in which μ_k is the transition dipole moment from $|g\rangle$ to $|k\rangle$ and $f(t)$ is the envelope of the pulse. A similar calculation used for the two-level model is used to solve the Schrödinger equation using Hamiltonians for the two-band model (Fig. 6.9).

6.4 Optical Detection Mechanism

The pump–probe transmission (or reflection measurement) can be treated as a third-order nonlinear optical response with a heterodyne detection. Here, we use the four-level model consisting of two electronic states ($|g\rangle$ and $|e\rangle$) and two phonon states ($|0\rangle$

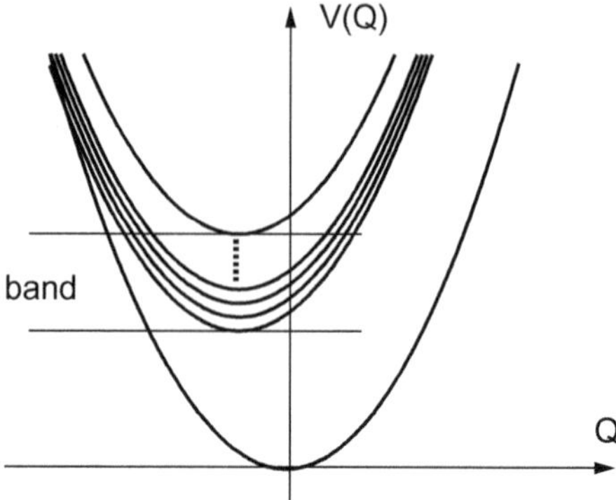

Fig. 6.9 A schematic of the potential for the two-band model with displaced harmonic oscillators

and $|1\rangle$) under transparent conditions. The pump pulse generates phonon polarization in the electronic ground state by ISRS and the second-order density operator is given by

$$\rho^{(2)}(t) = \alpha \left(\frac{\mu E_0}{\hbar}\right)^2 \int_{-\infty}^{t} dt' \int_{-\infty}^{t'} dt'' f(t'') f(t') \\ \times \left(e^{-i\omega(t-t'')} - e^{-i\omega(t-t')}\right) e^{-i\Delta(t'-t'')} |g, 1\rangle\langle g, 0| + H.c., \quad (6.121)$$

where $\Delta \equiv \varepsilon\hbar - \Omega$ is the detuning. By using a Gaussian pulse envelope with pulse width σ,

$$f(t) = \frac{1}{\sqrt{\pi}\sigma C} \exp\left(-\frac{t^2}{\sigma^2}\right), \quad (6.122)$$

where C is a dimensionless normalization factor ($\sqrt{\pi}\sigma C = 1$). Following the calculation of Sect. 6.1.5, we get

$$\rho^{(2)}(t) = i\alpha \frac{|\mu|^2 |E_0|^2}{\hbar^2} e^{-i\omega t} \frac{\omega}{\sqrt{2}\sigma C^2 \Delta^2} e^{-\sigma^2\omega^2/8} |g, 1\rangle\langle g, 0|, \quad (6.123)$$

and its Hermite conjugate. The expectation value of the excited coherent optical phonons $\langle Q\rangle$ can be calculated as

$$\langle Q(t)\rangle = Tr[Q\rho^{(2)}(t)] = Tr\left[\sqrt{\frac{\hbar}{2\omega}} (\hat{b} + \hat{b}^\dagger) \rho^{(2)}(t)\right] \\ = \sqrt{\frac{\hbar}{2\omega}} A |E_0|^2 \sin(\omega t), \quad (6.124)$$

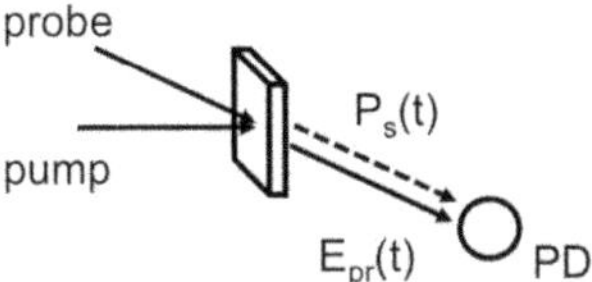

Fig. 6.10 Schematic of the pump–probe heterodyne detection of transient transmission. $E_{pr}(t)$ is the electronic field of the probe pulse and $P_s(t)$ is the induced polarization. The dotted arrow indicates dipole radiation from the induced polarization. PD is a photo detector

where, we set

$$A = \alpha \frac{\mu^2}{\hbar^2} \frac{\omega}{\sqrt{2}\sigma C^2 \Delta^2} e^{-\sigma^2\omega^2/8}. \tag{6.125}$$

The phonon amplitude is proportional to the intensity of the pump pulse. The displacement of atoms defined by the phonon amplitude can not be directly detected by the optical response, while it can be detected by using the X-ray diffraction. In the pump and probe protocol, the probe pulse induces electronic polarization $P_s(t)$, which causes dipole radiation. When a heterodyne detection of the transmitted probe pulse is investigated, the detection intensity $I_h(t)$ is

$$I_h(t) = \Omega l \times \mathrm{Im}(E_3^*(t) P_s(t)), \tag{6.126}$$

where l is the thickness of the sample [10], $E_3^*(t)$ is the strength of the electronic field of the probe pulse, and $\mathrm{Im}(A)$ is to get a imaginary part of A. Figure 6.10 shows a schematic drawing of the experimental configuration.

The probe pulse irradiates the sample at time t_p, which is a delay between the pump and probe pulses. There are eight pathways (Fig. 6.11) as given below:

$$\begin{aligned}
&\mathrm{path1} : |g, 1\rangle\langle g, 0| \to |e, 1\rangle\langle g, 0| \to |g, 0\rangle\langle g, 0| \\
&\mathrm{path2} : |g, 1\rangle\langle g, 0| \to |e, 0\rangle\langle g, 0| \to |g, 0\rangle\langle g, 0| \\
&\mathrm{path3} : |g, 1\rangle\langle g, 0| \to |g, 1\rangle\langle e, 1| \to |g, 1\rangle\langle g, 1| \\
&\mathrm{path4} : |g, 1\rangle\langle g, 0| \to |g, 1\rangle\langle e, 0| \to |g, 1\rangle\langle g, 1|
\end{aligned} \tag{6.127}$$

and their Hermite conjugates. The third-order density operator for the path 1 is obtained by

$$\begin{aligned}
\rho_1^{(3)}(t') &= iA|E_0|^2 \frac{i\mu}{\hbar} E_3 e^{-i\omega t_p} \\
&\times \int_{-\infty}^{t'} dt'' f_3(t'') e^{-i\omega t''} e^{-i\Omega t''} e^{-i(\varepsilon+\hbar\omega)(t'-t'')/\hbar} |e, 1\rangle\langle g, 0|,
\end{aligned} \tag{6.128}$$

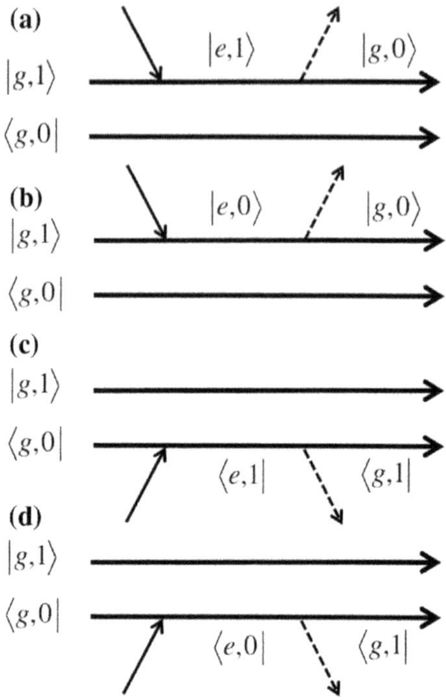

Fig. 6.11 The double-sided Feynman diagrams for a detection scheme in the pump and probe measurement

where $f_3(t')$ is the Gaussian pulse envelope of the probe pulse and t_p is the pump–probe delay. The polarization $(P(t))$ induced by the probe pulse is obtained using the polarization operator $\hat{P}^{op} \equiv \mu|g\rangle\langle e| + \mu^*|e\rangle\langle g|$ by $P(t) = Tr(\rho^{(3)}(t)\hat{P}^{op})$. The polarization for the path 1 is obtained as

$$\begin{aligned} P_1(t') &= \alpha A|E_0|^2\frac{\mu^2}{\hbar}E_3 e^{-i\omega t_p} \\ &\times \int_{-\infty}^{t'} \mathrm{d}t'' f_3(t'')e^{-i\omega t''}e^{-i\Omega t''}e^{-i(\varepsilon+\hbar\omega)(t'-t'')/\hbar}|e,1\rangle\langle g,0|, \end{aligned} \tag{6.129}$$

and the time-integrated intensity $(I_1(t_p))$ of the product between the probe light and polarization is as

$$\begin{aligned} I_1(t_p) &= \int_{-\infty}^{\infty} E_3 f_3^*(t')P_1(t')\mathrm{d}t' \\ &= \alpha A|E_0|^2\frac{\mu^2}{\hbar}|E_3|^2 e^{-i\omega t_p} \\ &\times \int_{-\infty}^{\infty}\mathrm{d}t'\int_{-\infty}^{t'}\mathrm{d}t'' f_3(t')e^{i\Omega t'} f_3(t'')e^{-i\omega t''} \\ &\times e^{-i\Omega t''}e^{-i(\varepsilon+\hbar\omega)(t'-t'')/\hbar}. \end{aligned} \tag{6.130}$$

For the Gaussian pulse, we get

$$
\begin{aligned}
I_1(t_p) &= \alpha A|E_0|^2 \frac{\mu^2}{\hbar\pi\sigma^2 C^2}|E_3|^2 e^{-i\omega t_p} \\
&\times \int_{-\infty}^{\infty} e^{-2s^2/\sigma^2} e^{-i\omega s} \int_0^{\infty} \mathrm{d}u e^{-u^2/(2\sigma^2)} e^{i(\Delta-\omega/2)u} \\
&\approx \frac{\mu^2|E_0|^2|E_3|^2}{\hbar\pi\sigma^2 C^2} e^{-i\omega t_p} \frac{\sqrt{\pi}}{\sqrt{2}\sigma} e^{-\sigma^2\omega^2/8} \frac{i\alpha A}{\Delta - \omega/2},
\end{aligned}
\tag{6.131}
$$

where $s = (t' + t'')/2$ and $u = t' - t''$. This is simply expressed by

$$
I_1(t_p) = iB|E_0|^2|E_3|^2 e^{i\omega t_p} \frac{1}{\Delta - \omega/2},
\tag{6.132}
$$

where

$$
B = \frac{\sqrt{\pi}\omega\alpha^2\mu^2}{2\sigma^2 C^4 \Delta^2 \hbar^3} e^{-\omega^2\sigma^2/4}.
\tag{6.133}
$$

A similar calculation shows that $I_4(t_p) = I_1(t_p)$ and

$$
I_2(t_p) = I_3(t_p) = iB|E_0|^2|E_3|^2 e^{-i\omega t_p} \frac{1}{\Delta + \omega/2}.
\tag{6.134}
$$

Thus, the total intensity is

$$
\begin{aligned}
I(t_p) &= \sum_i I_i(t_p) + H.c. = |E_0|^2|E_3|^2 e^{-i\omega t t_p} \frac{2iB\omega}{\Delta^2} + H.c. \\
&= |E_0|^2|E_3|^2 \frac{4B\omega}{\Delta^2} \sin(\omega t_p).
\end{aligned}
\tag{6.135}
$$

The results show that the change in transmission is proportional to the pump and probe intensity and proportional to the phonon amplitude $\langle Q(t')\rangle$ by comparing (6.124).

6.5 Summary

The generation processes of the coherent optical phonons were described quantum mechanically by using the two electronic levels and the displaced harmonic oscillators. Two generation processes (IA and ISRS) were investigated. Using the second-order perturbation, the mean value of the phonon amplitude was calculated for both the IA and ISRS processes. Detailed calculation was done with the Gaussian pulse for the Fourier-transform-limited case. The phonon oscillation is cosine-like and sine-like for the IA and ISRS processes, respectively. Both the IA and ISRS processes coexist in quantum mechanics. The ratio between the IA and ISRS processes is dependent on the pulse width and detuning. For the small deformation case ($\alpha \ll 1$), both the IA and ISRS processes are approximately expressed using the four-level model. The detection process of the EO sampling was also calculated using the same model.

References

1. Scholz, R., Pfeifer, T., Kurz, H.: Density-matrix theory of coherent phonon oscillations in germanium. Phys. Rev. B **24**, 16229–16236 (1993)
2. Kuznetsov, A.V., Stanton, C.J.: Theory of coherent phonon oscillations in semiconductors. Phys. Rev. Lett. **43**, 3243–3246 (1994)
3. Hu, X., Nori, F.: Quantum phonon optics: coherent and squeezed atomic displacement. Phys. Rev. B **53**, 2419–2424 (1996)
4. Merlin, R.: Generating coherent THz phonons with light pulses. Solid State Commun. **102**, 107–220 (1997)
5. Stevens, T.E., Kuhl, J., Merlin, R.: Coherent phonon generation and two stimulated Raman tensors. Phys. Rev. B **65**, 144304 (2002)
6. Nakamura, K.G., Shikano, Y., Kayanuma, Y.: Influence of pulse width and detuning on coherent phonon generation. Phys. Rev. **B92**, 144304 (2015)
7. Pollard, W.T., Fragnito, H.L., Bigot, J.-Y., Shank, C.V., Mathies, R.A.: Quantum-mechanical theory for 6 fs dynamic absorption spectroscopy. Chem. Phys. Lett. **168**, 239–245 (1990)
8. Banin, U., Bartana, A., Ruthman, S., Kosloff, R.: Impulsive excitation of coherent vibrational motion ground surface dynamics induced by intense short pulse. J. Chem. Phys. **101**, 8461–8481 (1994)
9. Cerullo, G., Manzoni, C.: Time domain vibrational spectroscopy: principle and experimental tools. In: De Silvestri, S., Cerullo, G., Lanzani, G. (eds.) Coherent Vibration Dynamics. CRC Press, Boca Raton (2008)
10. Mukamel, S.: Principles of Nonlinear Optical Spectroscopy. Oxford University Press, New York (1995)
11. Nakamura, K.G., Ohya, K., Takahashi, H., Tsuruta, T., Sasaki, H., Uozumi, S., Norimatsu, K., Kitajima, M., Shikano, Y., Kayanuma, Y.: Spectrally resolved detection in transient-reflectivity measurements of coherent optical phonons. Phys. Rev. **B94**, 024303 (2016)

Chapter 7
Coherent Control of Optical Phonons

This chapter describes the coherent control of optical phonons. We investigate, as an example, experiments and theoretical treatment of the coherent control of optical phonons in diamond, because it is one of the simplest systems. A four-level model consisting of two electronic states with two phonon states is used to calculate the coherent control of the optical phonons. Several selected examples of coherent control experiments of optical phonons are also presented.

7.1 Coherent Control

Coherent control is a technique to control quantum states using laser light and was originally developed for controlling chemical reactions using coherent two-photon process to assist chemical reactions via electronic excited states [1–3]. It has been performed for other physical properties, for example, electronic, vibrational, and rotational states of atoms and molecules, and electrons, excitons, spins, and phonons in condensed matter [4–8]. Here, we describe the coherent control of optical phonons using multiple ultrashort pulses.

7.2 Experiment of Coherent Control of Optical Phonons

The coherent control of optical phonons was first demonstrated on α-perylene molecular crystals at 5 K using multiple femtosecond pulses by Weiner et al. [9]. They irradiated a femtosecond pulse train consisting of more than 10 pulses, which was generated through a pulse-shaping technique, on the sample for exciting the optical phonons via the ISRS process in a transparent condition. They controlled the oscillation amplitude of the specific phonon mode to be approximately 5 THz owing to the frequency of the pulse train. A few years later, the coherent control of longitudinal optical (LO) phonon dynamics was demonstrated in a single crystal of a

K. Nakamura, *Quantum Phononics*, Springer Tracts in Modern Physics 282,
https://doi.org/10.1007/978-3-030-11924-9_7

semiconductor (gallium arsenide) in an opaque condition using a pair of femtosecond pulses at room temperature [10]. The first pump pulse generated 8.75 THz LO phonons by an ultrafast screening of the surface-space-charge field. The amplitude of the phonons was enhanced or reduced by the second pulse when the delay between the two pump pulses matched an integer or a half-integer multiple of the oscillation period through constructive or destructive interference, respectively. Similar coherent control experiments with double femtosecond pulses were performed on semimetal (bismuth) films [11] by a displacive excitation mechanism. The amplitude dependence on the delay between the two pump pulses was well explained using a combination of two damped oscillations. In addition to the amplitude control, the phase change in the controlled phonons was reported and its behavior deviated from a linear dependence, which was predicted by theoretical calculations. The coherent control of optical phonons is currently being used for selective excitation [9, 12, 13] and control of a structural transformation [14].

7.2.1 Coherent Control of Optical Phonons in Diamond

The coherent control of optical phonons in diamond is given as an example of coherent control using a pair of ultrashort optical pulses. The experimental setup used for this experiment is the same described in Fig. 5.1. The sub-10-fs optical pulse is separated into two pulses by using a beam splitter. One is used for the pump pulse and the other for the probe pulse. The pump pulse is directed to a Michelson-type interferometer (as shown in Fig. 7.1), and a pair of pump pulses (pump 1 and pump 2) is generated. Relative delay between pump 1 and pump 2 is precisely controlled by controlling difference between their optical path lengths.

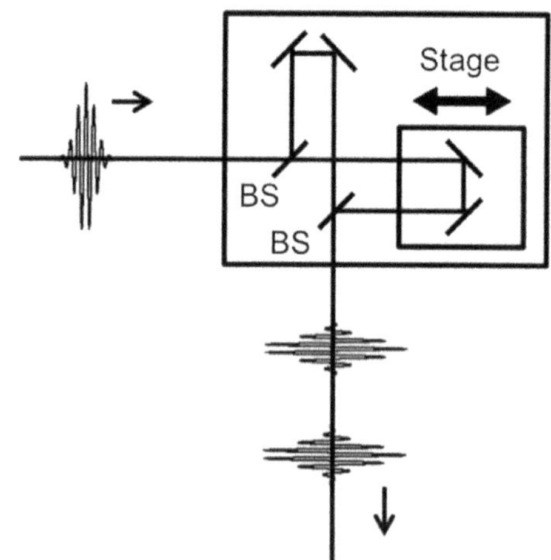

Fig. 7.1 Schematic of the Michelson-type interferometer, which is used to generate a pair of femtosecond pulses. BS is a beam splitter with 1:1 ratio

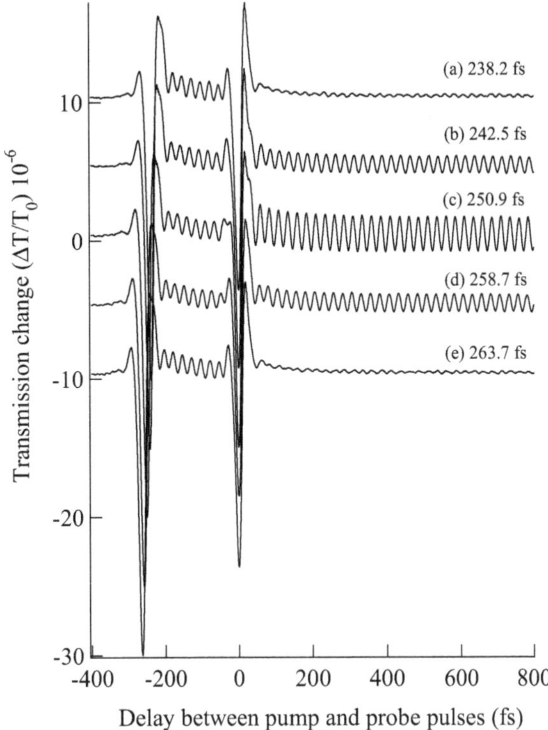

Fig. 7.2 Transient transmittance measurement in femtosecond laser pumped diamond using a pair of pump pulses for the pump–pump delay of 283.2 fs (**a**), 242.5 fs (**b**), 250.9 fs (**c**), 258.7 fs (**d**), and 263.7 fs (**e**). This figure is obtained by modifying Fig. 3 of the paper, H. Sasaki et al., Scientific Reports 8, 9609 (2018)

Figure 7.2 shows transient transmittance pumped by a pair of femtosecond optical pulses. The coherent oscillation due to optical phonons induced by the pump 1 is enhanced or suppressed by the pump 2 at delay 250.9 fs or 263.7 fs, respectively.

The oscillation amplitude and its initial phase after the pump 2 irradiation are shown in Fig. 7.3. The phonon amplitude is enhanced and suppressed at integer multiples and half-integer multiples of the vibrational period (25 fs).

7.3 Coherent Control Theory

7.3.1 Transparent Condition

At first, we consider the coherent control of optical phonons in diamond, which is described above. Phonon excitation processes by two pump pulses are expressed by the following double-sided Feynman diagrams (Fig. 7.4) for the second-order perturbation. The first diagram (a) indicates the phonons that are generated by the pump 1. The electronic excitation and relaxation occur with the pulse. In a similar way, the phonons are generated by the pump 2 in the second diagram (b). The third diagram indicates that the excitation occurs in the pump 1 and the relaxation occurs

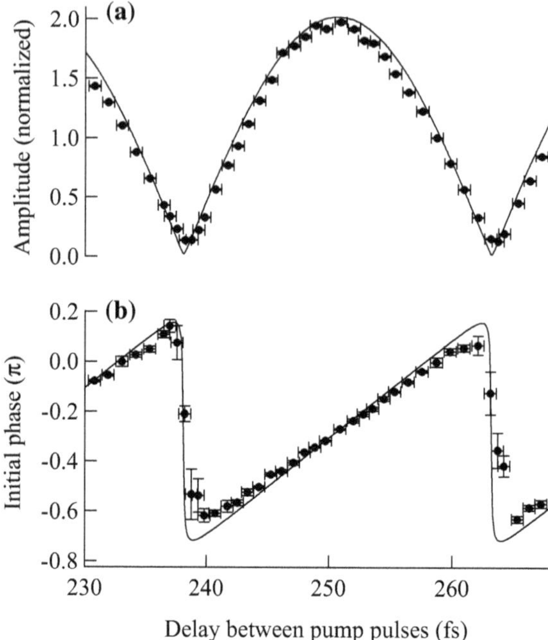

Fig. 7.3 The amplitude (**a**) and phase (**b**) of the controlled oscillation after the pump 2 against the pump–pump delay τ. The amplitude is normalized using that obtained after excitation by the pump 1. The solid circles are the experimental data, and the solid curves are obtained by the theoretical calculation. This figure is obtained by modifying Fig. 4 of the paper, H. Sasaki et al., Scientific Reports 8, 9609 (2018)

in the pump 2. The third process is important only when the two pump pulses overlap for the nonresonant condition (transparent condition). All processes are within the second-order perturbation. The density operator for the second diagram ($\rho_2^{(2)}$) is the same as the first one ($\rho_1^{(2)}$), which is shown in Chap. 6, except for the time delay:

$$\begin{aligned}\rho^{(2)}(t) &= \rho_1^{(2)} + \rho_2^{(2)} \\ &= iA\left(|E_1|^2 e^{-i\omega t} + |E_2|^2 e^{-i\omega(t-\tau)}\right)|g,1\rangle\langle g,0| + H.c.,\end{aligned} \tag{7.1}$$

where E_1 and E_2 are the electric field amplitude of the pump 1 and pump 2, τ is the delay between pump 1 and pump 2, and

$$A = \alpha \frac{\mu^2}{\hbar^2} \frac{\omega}{\sqrt{2}C^2\Delta^2} e^{-\sigma^2\omega^2/8}. \tag{7.2}$$

The mean value of the phonon amplitude $\langle Q(t)\rangle$ is calculated as

$$\langle Q(t)\rangle = \sqrt{\frac{\hbar}{2\omega}} A\left(|E_1|^2 \sin(\omega t) + |E_2|^2 \sin(\omega(t-\tau))\right). \tag{7.3}$$

The result shows that the phonon amplitude after the second pump pulse irradiation is a sum of two sinusoidal functions induced by each pulse. The phonon amplitude and initial phase at the timing of irradiation of pump 2 are calculated for diamond and shown in Fig. 7.3 as solid curves.

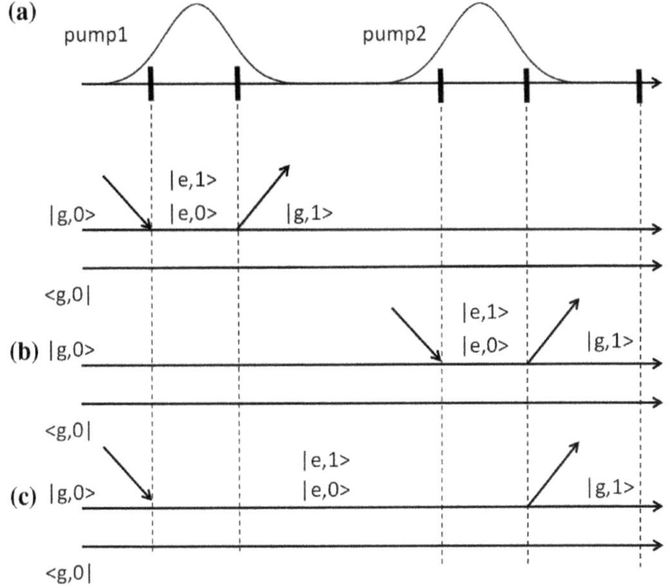

Fig. 7.4 Diagram of the impulsive stimulated Raman scattering paths. **a** passes through $|e, 1\rangle$ state and **b** passes through the $|e, 0\rangle$ state. Time flows from the felt to the right. The upper part of the figure shows an envelope of the optical pulse. Interaction between the optical pulse and the system occurs at time t″ and t′

It is worth noting that the above coherent control is due to constructive or destructive interference between phonon states excited by the pump 1 and pump 2. This is because we cannot distinguish which pulse excites the phonons while keeping the phonon coherence. The phonons are excited by the second-order perturbation, which interacts with the light twice in one pulse. This is different from the classical control of pendulum oscillations with two impacts by two pulses.

7.3.1.1 Short-Delay Region

When the delay between the pump 1 and pump 2 is very short, the third path represented as (c) in Fig. 7.4 is important. The density operator is calculated as

$$\rho_3^{(2)}(t,\tau) = -\alpha\left(\frac{\mu}{\hbar}\right)^2 e^{-i\Omega\tau}e^{-i\omega t}\int_{-\infty}^{t}\int_{-\infty}^{t'} dt''dt' f_1(t'')f_2(f'-\tau)$$
$$\times \exp\left(-i\frac{\varepsilon-\hbar\Omega}{\hbar}(t'-t'')\right)\left(e^{i\omega t''}-e^{i\omega t'}\right)|g,1\rangle\langle g,0| + H.c. \quad (7.4)$$

When the two pulses overlap, there is another transition path, in which the excitation and relaxation are induced by the pump 2 and pump 1, respectively. The density operator of this path is

$$\rho_4^{(2)}(t,\tau) = -\alpha\left(\frac{\mu}{\hbar}\right)^2 e^{i\Omega\tau}e^{-i\omega t}\int_{-\infty}^{t}\int_{-\infty}^{t'} dt''dt' f_1(t')f_2(f''-\tau)$$
$$\times\exp\left(-i\frac{\varepsilon-\hbar\Omega}{\hbar}(t'-t'')\right)\left(e^{i\omega t''}-e^{i\omega t'}\right)|g,1\rangle\langle g,0| + H.c. \quad (7.5)$$

With $\rho_3^{(2)}(t,\tau)$ and $\rho_4^{(2)}(t,\tau)$ included, the expectation value of the phonon amplitude should be modified with the optical frequency Ω. This oscillation can be observed when the delay between two pump pulses is controlled with a step shorter than the optical cycle.

7.3.2 Opaque Condition

In the opaque conditions, the impulsive absorption paths are possible in addition to the ISRS paths described above. The final state should be $|e,1\rangle\langle e,0|$ or $|e,0\rangle\langle e,1|$.

The paths (a) and (b) are the same as the single pulse excitation described in Chap. 6. The density operators for the path (c) and (d) in Fig. 7.5 are

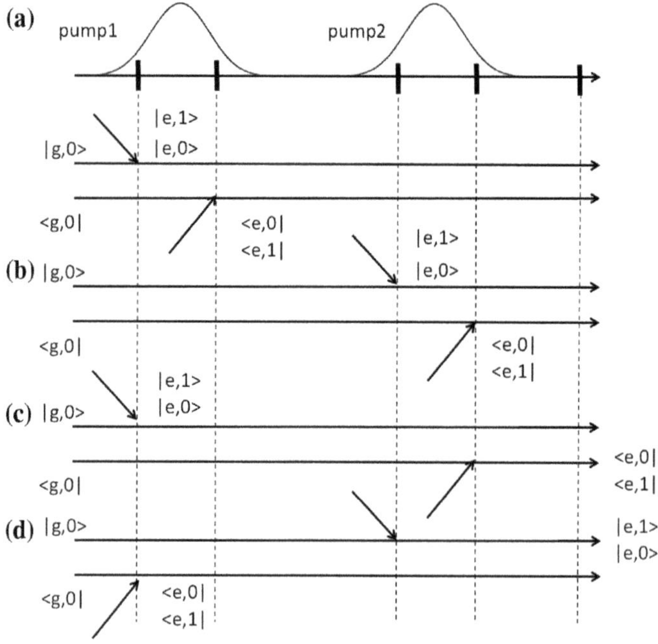

Fig. 7.5 Diagrams of the coherent control of optical phonons with the impulsive absorption paths. Transition occurs by only the pump 1 (**a**) or the pump 2 (**b**). (**c**) and (**d**) represent the transition occurring by both the pump 1 and pump 2. Time flows from the felt to the right. The upper part of the figure shows an envelope of the optical pulse. Interaction between the optical pulse and the system occurs at time t″ and t′

$$\rho_3^{(2)}(t,\tau) = \alpha\left(\frac{\mu}{\hbar}\right)^2 e^{-i\Omega\tau}\int_{-\infty}^{t}\int_{-\infty}^{t} \mathrm{d}t''\mathrm{d}t' f_1(t'')f_2(f'-\tau)$$
$$\times \exp\left(-i\frac{\varepsilon-\hbar\Omega}{\hbar}(t'-t'')\right)$$
$$\times\left(e^{-i\omega(t-t'')}|e,1\rangle\langle e,0| + e^{i\omega(t-t')}|e,0\rangle\langle e,1|\right), \tag{7.6}$$

and

$$\rho_4^{(2)}(t,\tau) = \alpha\left(\frac{\mu}{\hbar}\right)^2 e^{i\Omega\tau}\int_{-\infty}^{t}\int_{-\infty}^{t} \mathrm{d}t''\mathrm{d}t' f_1(t'')f_2(f'-\tau)$$
$$\times \exp\left(-i\frac{\varepsilon-\hbar\Omega}{\hbar}(t'-t'')\right)$$
$$\times\left(e^{-i\omega(t-t')}|e,1\rangle\langle e,0| + e^{i\omega(t-t'')}|e,0\rangle\langle e,1|\right). \tag{7.7}$$

This shows that the phonon amplitude is also modulated by the optical frequency.

7.4 Selected Examples

There are many applications of the coherent control of optical phonons such as the mode-selective excitation and phase-transition control. We describe several examples of the coherent control experiments of the coherent phonons.

7.4.1 *Mode-Selective Excitation*

$YBa_2Cu_3O_{7-\delta}$ is a typical example of a high-temperature transition superconductor and has two optical phonons, Ba–O and Cu–O modes, at frequencies of 3.4 and 4.3 THz, respectively. Both phonon modes are excited by an ultrashort pulse. By using a double-pulse excitation, we can selectively excite either the Ba–O mode phonons or the Cu–O mode phonons, because their vibrational periods are different: 294.1 and 232.6 fs for the Ba–O and Cu–O modes, respectively. Takahashi et al. [12] selectively excited the phonon modes using a pair of femtosecond pulses and showed that the Ba–O and Cu–O modes are strongly suppressed at the double-pulse separation time of 135.0 and 108.5 fs, which are close to a half-integer multiple of the phonon periods. At the separation time of 135.0 fs, only the Cu–O mode oscillation was observed. The phonon amplitude of the Ba–O and Cu–O modes are enhanced at separation time of 270 and 217 fs, respectively.

Katsuki et al. [13] demonstrated all-optical control and visualization of ultrafast two-dimensional atomic motions in a single crystal of bismuth using a pair of chirped femtosecond pulses. Bismuth has two optical phonons A_{1g} and E_g modes with different frequencies of 3 and 2 THz, respectively. Both phonon modes are excited by an ultrashort optical pulse and detected via the EO sampling technique. In the A_{1g} and E_g modes, the Bi atoms oscillate in the longitudinal and lateral directions, when the (0001) surface is used. The relative intensity ratio between the A_{1g} and E_g modes, which corresponds to the displacement direction of atoms in the two-dimensional space, was controlled by controlling the delay between two pump pulses. The reflectivity change shows beat structures, which are produced by overlapping two oscillations, whose temporal evolution changes drastically as they change the delay between two pump pulses on the attosecond timescale. They mapped these beat structures into a two-dimensional space atomic displacement with the density functional theory calculation of the reflectivity change.

7.4.2 Mode Coupling

Not only the LO phonons ($\omega_L/2\pi = 8.7$ THz) but also the optical phonon–plasmon coupled (LOPC) mode oscillations ($\omega_-/2\pi = 7.7$ THz) are coherently controlled by using a pair of pump pulses at delays between 315 and 345 fs [15]. In n-GaAs, the oscillation amplitudes of the LO phonons and LOPC mode were harmonically modulated according to their frequencies by the pump–probe delay. On the other hand, in p-GaAs, the oscillation amplitude of the LOPC mode was modulated in phase with that of the LO phonons at the period of LO phonons. They analyzed the results using a phenomenological model by assuming that the LOPC formation is delayed from the LO-phonon excitation. The lifetime of the LOPC plays a key role in understanding different results in two samples. The lifetime of the LOPC is ~200 fs in p-GaAs, and ~800 fs in n-GaAs. Therefore, when the second pump pulse arrives at the sample, the LOPC excited by the first pump pulse has already decayed significantly in p-GaAs. Thus, the LOPC excited by the second pump pulse has nearly no interference with that excited by the first one, and the dynamics are determined mainly by the interference of LO phonons. The excited LO phonons cause the phonon–plasmon coupling. The delay time of the LOPC formation was estimated to be 100 fs for n-GaAs and 120 fs for p-GaAs. In addition to the control of phonon amplitudes, they reported that the lifetime of the LOPC was coherently controlled, while that of LO phonons was independent of the pump–pulse delay.

7.4.3 Controlling Phase Transition

$Ge_2Sb_2Te_5$ (GST) is a chalcogenide phase-change material, in which phase transformation between the amorphous and crystal phase is optically induced. Makino et al. [16] demonstrated that the phase change from amorphous into crystalline states is manipulated by controlling atomic motions through selectively exciting vibrational modes that involve Ge atom using a pair of femtosecond pulses. The GST superlattice sample was irradiated by the two pump pulses with delay of 276 fs, which is close to the time period of the local A_1 mode of the octahedral $GeTe_6$ (in crystalline phase). The pump pulses coherently enhanced the vibrational amplitude toward the crystalline phase.

7.5 Summary

We described the experiments and theoretical treatments of the coherent control of optical phonons in diamond, because it is one of the simplest systems. A four-level model consisting of two electronic states with two phonon states is used to calculate the coherent control of the phonon amplitude. Several selected examples of coherent control experiments of optical phonons are also presented.

References

1. Tannor, D.J., Rice, S.A.: Control of selectivity of chemical reaction via control of wave packet evolution. J. Chem. Phys. **83**, 5013 (1985)
2. Brumer, P., Shapiro, M.: Control of unimolecular reactions using coherent light. Chem. Phys. Lett. **126**, 541 (1986)
3. Tannor, D.J., Kosloff, R., Rice, S.A.: Coherent pulse sequence induce control of selectivity of reactions: exact quantum mechanical calculations. J. Chem. Phys. **85**, 5805 (1986)
4. Ramsay, A.J.: A review of the coherent optical control of the exciton and spin states of semiconductor quantum dots. Semicond. Sci. Techol. **25**, 103001 (2010). and references therein
5. Katsuki, H., Takei, N., Sommer, C., Ohmori, K.: Ultrafast coherent control of condensed matter with attosecond precison. Acc. Chem. Res. **51**, 1174–1184 (2018)
6. Okano, Y., Katsuki, H., Nakagawa, Y., Takahashi, H., Nakamura, K.G., Ohmori, K.: Optical manipulation of coherent phonons in superconducting $YBa_2Cu_4O_{7-\delta}$ thin films. Faraday Discuss. **153**, 375–382 (2011)
7. Cheng, Y.-H., Gao, F.Y., Teitelbaum, S.W., Nelson, K.A.: Coherent control of optical phonons in bismuth. Phys. Rev. B **96**, 134302 (2017)
8. Hu, J., Igarashi, K., Sasagawa, T., Nakamura, K.G., Misochko, O.V.: Femtosecond study of A_{1g} phonons in the strong 3D topological insulators: From pump-probe to coherent control. Appl. Phys. Lett. **112**, 031901 (2018)
9. Weiner, A.M., Leaird, D.E., Wiederrecht, G.P., Nelson, K.A.: Femtosecond multiple-pulse impulsive stimulated Raman scattering spectroscopy. J. Opt. Soc. Am. B **8**, 1264 (1991)
10. Dekorsky, T., Kütt, W., Pfeifer, T., Kurz, H.: Coherent Control of LO-Phonon Dynamics in Opaque Semiconductors by Femtosecond Laser Pulses. Europhys. Lett. **23**, 223 (1993)
11. Hase, M., Mizoguchi, K., Harima, H., Nakashima, S., Tani, M., Sakai, K., Hangyo, M.: Optical control of coherent optical phonons in bismuth films. Appl. Phys. Lett. **69**, 2474 (1996)
12. Takahashi, H., Kato, K., Nakano, H., Kitajima, M., Ohmori, K., Nakamura, K.G.: Optical control and mode selective excitation of coherent phonons in $YBa_2Cu_3O_{7-\delta}$. Solid State Commun. **149**, 1955 (2009)
13. Katsuki, H., Delagnes, J.C., Hosaka, K., Ishioka, K., Chiba, H., Zijlstra, E.S., Garcia, M.E., Takahashi, H., Watanabe, K., Kitajima, M., Matsumoto, Y., Nakamura, K.G., Ohmori, K.: All-optical control and visualization of ultrafast two-dimensional atomic motions in a single crystal of bismuth. Nat. Commun. **4**, 2801 (2013)

14. Hase, M., Fons, P., Mitrofanov, K., Kolobov, A.V., Tominaga, J.: Femtosecond structural transformation of phase-change materials far from equilibrium monitored by coherent phonons. Nat. Commun. **6**, 8367 (2015)
15. Hu, J., Misochko, O.V., Goto, A., Nakamura, K.G.: Delayed formation of coherent LO phonon-plasmon coupled modes in n- and p-type GaAs measured using a femtoseocnd coherent control technique. Phys. Rev. B **86**, 235145 (2012)
16. Makino, K., Tominaga, J., Hase, M.: Ultrafast optical manipulation of atomic arrangements in chalcogenide alloy memory materials. Opt. Express **19**, 138475 (2011)

Appendix A
Linear Diatomic Chain: Normal Coordinate and Hamiltonian

Here, we describe the normal coordinates of the linear diatomic chain and diagonalize its Hamiltonian.[1]

A.1 Normal Coordinates

The displacement of atom κ with the wavenumber k is expressed by

$$q^l_{\kappa,j}(k) = \frac{1}{2\sqrt{Nm_\kappa}}\left(A_{0,j}(k)a_{\kappa,j}(k)e^{ikx^l_\kappa - i\omega_j(k)t} + c.c.\right). \tag{A.1}$$

When we set $q > 0$, the displacement of the atomκ at the lth cell is expressed by

$$\begin{aligned} q^l_\kappa &= \frac{1}{2\sqrt{Nm_\kappa}}\sum_{j,k\geq 0}\left(A_{0,j}(k)a_\kappa(k)e^{ikx^l_\kappa - i\omega_j(k)t} + c.c.\right) \\ &+ \frac{1}{2\sqrt{Nm_\kappa}}\sum_{j,k>0}\left(A_{0,j}(-k)a_\kappa(-k)e^{-ikx^l_\kappa - i\omega_j(-k)t} + c.c.\right). \end{aligned} \tag{A.2}$$

This solution is more generally expressed in terms of N complex normal coordinate Q^k_j as

$$q^l_\kappa = \frac{1}{\sqrt{Nm_\kappa}}\left(\sum_{j,k\geq 0} a_{\kappa,j}(k)e^{ikx^l_\kappa}Q^k_j + \sum_{j,k>0} a_{\kappa,j}(-k)e^{-ikx^l_\kappa}Q^{-k}_j\right), \tag{A.3}$$

where

[1] We use the method described in [1].

K. Nakamura, *Quantum Phononics*, Springer Tracts in Modern Physics 282,
https://doi.org/10.1007/978-3-030-11924-9

$$
\begin{aligned}
Q_j^k &= \frac{1}{2}\left(A_{0,j}(k)e^{-i\omega_j(k)t} + A_{0,j}(k)e^{i\omega_j(-k)t}\right) \\
&= \frac{1}{2}\left(A_{0,j}(k)e^{-i\omega_j(k)t} + A_{0,j}^*(-k)e^{i\omega_j(k)t}\right).
\end{aligned} \tag{A.4}
$$

Using $a_{\kappa,j}^*(-k) = a_{\kappa,j}(k)$, we calculate q_κ^{l*} as

$$
\begin{aligned}
q_\kappa^{l*} &= \frac{1}{\sqrt{Nm_\kappa}}\left(\sum_{j,k\geq 0} a_{\kappa,j}^*(k)e^{-ikx_\kappa^l}Q_j^{k*} + \sum_{j,k>0} a_{\kappa,j}^*(-k)e^{ikx_\kappa^l}Q_j^{-k*}\right) \\
&= \frac{1}{\sqrt{Nm_\kappa}}\left(\sum_{j,k\geq 0} a_{\kappa,j}^*(k)e^{-ikx_\kappa^l}Q_j^{k*} + \sum_{j,k>0} a_{\kappa,j}^*(-k)e^{ikx_\kappa^l}Q_j^{-k*}\right).
\end{aligned} \tag{A.5}
$$

Since the displacement q_κ^l is real, $q_\kappa^{l*} = q_\kappa^l$. Then, we get $Q_j^{k*} = Q_j^{-k}$. One of the most important merits of using the normal coordinates is that the Hamiltonian is diagonalized by using the normal coordinate (as shown in the next subsection).

A.2 Diagonalizing Hamiltonian

Hamiltonian H is a sum of kinetic energy T and potential energy V: $H = T + V$. The potential energy is

$$
T = \sum_{l,\kappa} \frac{\left(p_\kappa^l\right)^2}{2m_\kappa}, \tag{A.6}
$$

where p_κ^l is a momentum and defined by

$$
\begin{aligned}
p_\kappa^l &= m_\kappa \frac{\mathrm{d}}{\mathrm{d}t} q_\kappa^l = m_\kappa \dot{q}_\kappa^l \\
&= \sqrt{\frac{m_\kappa}{N}} \sum_{k,j} a_{\kappa,j}(k)\exp(ikx_\kappa^l)\dot{Q}_j(k).
\end{aligned} \tag{A.7}
$$

Using the normal coordinates, T is

$$
\begin{aligned}
T = \frac{1}{2N}\sum_{l,\kappa}&\left(\sum_{k,j} a_{\kappa,j}(k)\exp(ikx_\kappa^l)\dot{Q}_j(k)\right) \\
&\times\left(\sum_{k',j'} a_{\kappa,j'}(k')\exp(ik'x_\kappa^l)\dot{Q}_{j'}(k')\right)
\end{aligned}
$$

$$= \frac{1}{2N}\sum_{l,\kappa} m_\kappa \sum_{k,k',j,j'} a_{\kappa,j}(k)a_{\kappa,j'}(k')\exp(i(k+k')la)$$
$$\times \exp(i(k+k')x_\kappa)\dot{Q}_j(k)\dot{Q}_{j'}(k'). \tag{A.8}$$

This is not zero for $k' = -k$ condition because $\sum_l \exp(i(k-k')la) = N\Delta(k-k')$. Then, we get

$$\begin{aligned} T &= \frac{1}{2}\sum_\kappa \sum_{k,j,j'} a_{\kappa,j}(k)a_{\kappa,j'}(-k)\dot{Q}_j(k)\dot{Q}_{j'}(-k) \\ &= \frac{1}{2}\sum_\kappa \sum_{k,j,j'} a_{\kappa,j}(k)a^*_{\kappa,j'}(k)\dot{Q}_j(k)\dot{Q}^*_{j'}(k) \\ &= \frac{1}{2}\sum_{k,j} \dot{Q}_j(k)\dot{Q}^*_{j'}(k), \end{aligned} \tag{A.9}$$

where we use $\sum_\kappa a^*_{\kappa,j}(k)a_{\kappa,j'}(k) = \delta_{j,j'}$.

Next we calculate the potential energy V using (4.38):

$$V = \frac{1}{2}\left(fS + gS'\right), \tag{A.10}$$

where

$$\begin{aligned} S &= \sum_l \left(q_2^l - q_1^l\right)^2 \\ S' &= \sum_l \left(q_1^{l+1} - q_2^l\right)^2. \end{aligned} \tag{A.11}$$

Using the displacements described with normal coordinates

$$\begin{aligned} q_2^l &= \sum_{q,j} \frac{1}{\sqrt{Nm_2}} a_{2,j}(k)\exp(ik(la + x_2))Q_j(k) \\ q_1^l &= \sum_{q,j} \frac{1}{\sqrt{Nm_2}} a_{1,j}(k)\exp(ik(la + x_1))Q_j(k) \\ q_1^{l+1} &= \sum_{q,j} \frac{1}{\sqrt{Nm_2}} a_{1,j}(k)\exp(ik(la + x_1 + a))Q_j(k), \end{aligned} \tag{A.12}$$

we get

$$
\begin{aligned}
\sum_l \left(q_2^l\right)^2 &= \sum_l \left(\sum_{k,j} \frac{1}{\sqrt{Nm_2}} a_{2,j}(k) \exp(ik(la + x_2)) Q_j(k) \right) \\
&\times \left(\sum_{k',j'} \frac{1}{\sqrt{Nm_2}} a_{2,j'}(k') \exp(ik'(la + x_2)) Q_{j'}(k') \right) \\
&= \frac{1}{Nm_2} \sum_l \sum_{k,k',j,j'} a_{2,j}(k) a_{2,j'}(k') \exp(i(k+k')la) \\
&\times \exp(i(k+k')x_2) Q_j(k) Q_{j'}(k') \\
&= \frac{1}{m_2} \sum_{k,j,j'} a_{2,j}(k) a_{2,j'}(-k) Q_j(k) Q_{j'}(-k) \\
&= \frac{1}{m_2} \sum_{k,j,j'} a_{2,j}(k) a^*_{2,j'}(k) Q_j(k) Q^*_{j'}(k).
\end{aligned} \tag{A.13}
$$

Similar calculation gives

$$
\sum_l \left(q_1^l\right)^2 = \frac{1}{m_1} \sum_{k,j,j'} a_{1,j}(k) a^*_{1,j'}(k) Q_j(k) Q^*_{j'}(k), \tag{A.14}
$$

$$
\begin{aligned}
\sum_l \left(q_1^l\right)\left(q_2^l\right) &= \sum_l \left(\sum_{k,j} \frac{1}{\sqrt{Nm_1}} a_{1,j}(k) \exp(ik(la + x_1)) Q_j(k) \right) \\
&\times \left(\sum_{k',j'} \frac{1}{\sqrt{Nm_2}} a_{2,j'}(k') \exp(ik'(la + x_2)) Q_{j'}(k') \right) \\
&= \frac{1}{\sqrt{m_1 m_2}} \sum_{k,j,j'} a_{1,j}(k) a^*_{2,j'}(k) Q_j(k) Q^*_{j'}(k) \exp(ik(x_1 - x_2)) \\
&= \frac{1}{\sqrt{m_1 m_2}} \sum_{k,j,j'} a_{1,j}(k) a^*_{2,j'}(k) Q_j(k) Q^*_{j'}(k) \psi_{12},
\end{aligned} \tag{A.15}
$$

where $\psi_{12} \equiv \exp(ik(x_1 - x_2))$. Similarly, we get

$$
\sum_l \left(q_2^l\right)\left(q_1^l\right) = \sum_l \left(\sum_{k,j} \frac{1}{\sqrt{Nm_2}} a_{1,j}(k) \exp(ik(la + x_2)) Q_j(k) \right)
$$

$$\times \left(\sum_{k',j'} \frac{1}{\sqrt{Nm_1}} a_{1,j'}(k') \exp(ik'(la + x_1)) Q_{j'}(k') \right)$$
$$= \frac{1}{\sqrt{m_1 m_2}} \sum_{k,j,j'} a^*_{1,j}(k) a_{2,j'}(k) Q_j(k) Q^*_{j'}(k) \psi^*_{12}. \tag{A.16}$$

Using these equations, we get

$$S = \sum_l \left(\left(q_2^l\right)^2 + \left(q_2^l\right)\left(q_1^l\right) + \left(q_1^l\right)\left(q_2^l\right) + \left(q_1^l\right)^2 \right)$$
$$= \sum_{k,j,j'} Q_j(k) Q^*_{j'}(k) \Big\{ \frac{1}{m_1} a_{1,j}(k) a^*_{1,j'}(k) + \frac{1}{m_2} a_{2,j}(k) a^*_{2,j'}(k)$$
$$- \frac{1}{\sqrt{m_1 m_2}} \left(a_{1,j}(k) a^*_{2,j'}(k) \psi_{12} + a^*_{1,j'}(k) a_{2,j}(k) \psi^*_{12} \right) \Big\}. \tag{A.17}$$

The S' term is also calculated in the similar way:

$$S' = \sum_{k,j,j'} Q_j(k) Q^*_{j'}(k) \Big\{ \frac{1}{m_1} a_{1,j}(k) a^*_{1,j'}(k) + \frac{1}{m_2} a_{2,j}(k) a^*_{2,j'}(k)$$
$$- \frac{1}{\sqrt{m_1 m_2}} \left(a_{1,j}(k) a^*_{2,j'}(k) \Psi_{12} + a^*_{1,j'}(k) a_{2,j}(k) \Psi^*_{12} \right) \Big\}, \tag{A.18}$$

where ψ_{12} in S is replaced by Ψ_{12}, which is defined by $\Psi_{12} \equiv \exp(ik(x_1 - x_2 + a))$. The potential energy V is obtained as

$$V = \frac{1}{2} \sum_{k,j,j'} Q_j(k) Q^*_{j'}(k) \Big\{ \frac{f+g}{m_1} a_{1,j}(k) a^*_{1,j'}(k) + \frac{f+g}{m_2} a_{2,j}(k) a^*_{2,j'}(k)$$
$$- \frac{1}{\sqrt{m_1 m_2}} \Big(a_{1,j}(k) a^*_{2,j'}(k) \left(f\psi_{12} + g\Psi_{12} \right) + a^*_{1,j'}(k) a_{2,j}(k) \left(f\psi^*_{12} + g\Psi^*_{12} \right) \Big) \Big\}. \tag{A.19}$$

Using (4.48), the terms are replaced as

$$- \frac{1}{\sqrt{m_1 m_2}} \left(f\psi_{12} + g\Psi_{12} \right) = D^*_{12}(k) = D_{21}(k)$$
$$- \frac{1}{\sqrt{m_1 m_2}} \left(f\psi^*_{12} + g\Psi^*_{12} \right) = D_{12}(k) = D^*_{21}(k), \tag{A.20}$$

and we get

$$
\begin{aligned}
V &= \frac{1}{2}\sum_{k,j,j'} Q_j(k)Q^*_{j'}(k)\Big\{D_{11}(k)a_{1,j}(k)a^*_{1,j'}(k) + D_{22}(k)a_{2,j}(k)a^*_{2,j'}(k) \\
&+ D_{21}(k)a_{1,j}(k)a^*_{2,j'}(k) + D_{12}(k)a^*_{1,j'}(k)a_{2,j}(k)\Big\} \\
&= \frac{1}{2}\sum_{k,j,j'} Q_j(k)Q^*_{j'}(k)\Big\{\left(D_{11}(k)a_{1,j}(k) + D_{12}(k)a_{2,j}(k)\right)a^*_{1,j'}(k) \\
&+ \left(D_{22}(k)a_{2,j}(k) + D_{21}(k)a^*_{1,j}(k)\right)a^*_{2,j'}(k)\Big\}.
\end{aligned}
\tag{A.21}
$$

Using (4.50), the equation is simplified as

$$
\begin{aligned}
V &= \frac{1}{2}\sum_{k,j,j'} Q_j(k)Q^*_{j'}(k)\Big\{\omega_j^2 a_{1,j}(k)a^*_{1,j'}(k) + \omega_j^2 a_{2,j}(k)a^*_{2,j'}(k)\Big\} \\
&= \frac{1}{2}\sum_{k,j} Q_j(k)Q^*_j(k)\omega_j^2.
\end{aligned}
\tag{A.22}
$$

The Lagrangian is expressed using (A.8) and (A.22)

$$
L = T - V = \frac{1}{2}\sum_{k,j}\Big\{\dot{Q}_j(k)\dot{Q}^*_j(k) - Q_j(k)Q^*_j(k)\omega_j^2\Big\},
\tag{A.23}
$$

and the generalized momentum $P_j(k)$ conjugate to the normal coordinate $Q_j(k)$ is defined from the Lagrangian:

$$
P_j(k) = \frac{\partial L}{\partial \dot{Q}_j(k)} = \dot{Q}_j(k).
\tag{A.24}
$$

The Hamiltonian expressed using the normal coordinates and the conjugated momentum is

$$
H = T + V = \frac{1}{2}\sum_{k,j}\Big\{P_j(k)P^*_j(k) + Q_j(k)Q^*_j(k)\omega_j^2\Big\}.
\tag{A.25}
$$

Appendix B
Ultrashort Laser Technology

An ultrashort optical pulse is required to excite coherent optical phonons, because its pulse width should be shorter than vibrational period (lower than several hundred femtoseconds). Then, we show a brief summary of ultrashort laser technology.

B.1 Pulse and Spectrum Width

An ultrashort pulse consists of many continuous waves with different frequencies. Let us consider a simple Gaussian light pulse, which does not have time-dependent phase,

$$E(t) = E_0 e^{-t^2/\sigma^2} e^{i\omega_0 t}, \tag{B.1}$$

where E_0 is the amplitude of the electric field, σ is the pulse width, and ω_0 is the angular frequency of the light. The spectrum of the pulse is calculated by the Fourier transformation of the pulse as

$$E(\omega) = \frac{1}{\sqrt{2\pi}} \int_{-\infty}^{\infty} E(t) e^{-i\omega t} \mathrm{d}t = \frac{\sigma}{\sqrt{2}} E_0 e^{-(\omega-\omega_0)^2\sigma^2/4}. \tag{B.2}$$

The spectrum has a Gaussian form at the center angular frequency of ω_0. The intensities of the pulse and the spectrum are defined by $I(t) = E(t)E^*(t)$ and $I(\omega) = E(\omega)E^*(\omega)$, respectively. The widths (full width of half maximum) of the pulse and the spectrum are obtained as $\Delta t = \sigma\sqrt{2\ln 2}$ and $\Delta\omega = 2\pi \times \Delta\nu = 4\ln 2/\sigma$, respectively [2]. The relation between the pulse width and the frequency is

$$\Delta t \Delta \nu = \frac{2\ln 2}{\pi} \approx 0.441. \tag{B.3}$$

K. Nakamura, *Quantum Phononics*, Springer Tracts in Modern Physics 282,
https://doi.org/10.1007/978-3-030-11924-9

In more general case, the pulse includes a time-dependent phase term ($e^{i\theta(t)}$ and $e^{-i\theta(\omega)}$) and $\Delta t \Delta \nu$ is bigger than 0.441. The pulse satisfying (B.3) is called the Fourier-transform-limited pulse.

When the Fourier-transform-limited pulse passes through optical materials such as a lens and a beamsplitter, the pulse gets a frequency-dependent phase change [3]. The pulse after passing the optics is expressed as

$$E'(t) = \frac{1}{\sqrt{2\pi}} \int_{\infty}^{\infty} E(\omega) e^{i\theta(\omega)} e^{i\omega t} \mathrm{d}\omega. \tag{B.4}$$

The phase change is expressed by the Taylor expansion around ω_0

$$\begin{aligned}\theta(\omega) = \theta_0 &+ \left.\frac{\mathrm{d}\theta}{\mathrm{d}\omega}\right|_{\omega_0} (\omega - \omega_0) + \left.\frac{1}{2}\frac{\mathrm{d}^2\theta}{\mathrm{d}\omega^2}\right|_{\omega_0} (\omega - \omega_0)^2 \\ &+ \left.\frac{1}{6}\frac{\mathrm{d}^3\theta}{\mathrm{d}\omega^3}\right|_{\omega_0} (\omega - \omega_0)^3 + \cdots,\end{aligned} \tag{B.5}$$

where θ_0 is a constant, the second term is a group velocity delay, and the third term is a group velocity dispersion. If we consider only the third term using

$$\delta \equiv \left.\frac{1}{2}\frac{\mathrm{d}^2\theta}{\mathrm{d}\omega^2}\right|_{\omega_0}, \tag{B.6}$$

the electric field amplitude is given by

$$\begin{aligned}E'(t) &= \frac{1}{\sqrt{2\pi}} \int_{\infty}^{\infty} E(\omega) e^{i\theta(\omega)} e^{i\omega t} \mathrm{d}\omega \\ &= \frac{E_0\sigma}{2\sqrt{\pi}} \int_{\infty}^{\infty} \exp\left(-\frac{(\omega - \omega_0)^2\sigma^2}{4}\right) \exp(i\delta(\omega - \omega_0)^2) \exp(i\omega t) \mathrm{d}\omega \\ &= \frac{E_0\sigma}{2\sqrt{\pi}} e^{i\omega_0 t} \int_{\infty}^{\infty} \exp\left(-\left(\frac{\sigma^2}{4} - i\delta\right) u^2 + itu\right) \mathrm{d}u,\end{aligned} \tag{B.7}$$

where we set $u = \omega - \omega_0$. Using the Gaussian integral formula, we get

$$\begin{aligned}E'(t) &= \frac{E_0\sigma}{2\sqrt{\pi}} e^{i\omega_0 t} \sqrt{\frac{4\pi}{\sigma^2 - i4\delta}} \exp\left(\frac{-(\omega - \omega_0)^2}{\sigma^2 - i4\delta}\right) \\ &= \frac{E_0\sigma}{2\sqrt{\pi}} e^{i\omega_0 t} \sqrt{\frac{4\pi}{\sigma^2 - i4\delta}} \exp\left(\frac{-i4\delta t^2}{\sigma^4 + 16\delta^2}\right) \\ &\times \exp\left(-\frac{t^2}{\sigma^2 + 16\delta^2/\sigma^2}\right).\end{aligned} \tag{B.8}$$

The result indicates that the frequency is dependent on time and the pulse width is larger than that of the original pulse. This effect is called "chirp".

B.2 Optical Interference

The pulse width of femtosecond optical pulses cannot be measured directly by an electronic devise, because it has not enough time resolution. The pulse shape is monitored via optical interference for a Fourier transfer limited pulse [4]. A femtosecond pulse is divided into two pulses (pulse 1 and pulse 2) by a beam splitter. The pump 1 and pump 2 are mixed after passing the different optical paths and detected by a slow photodetector. The electric fields of the pump 1 and pump 2 are

$$\begin{aligned} E_1(t) &= E_0 \exp\left(-\frac{t^2}{\sigma^2}\right) e^{-i\omega t} \\ E_2(t) &= E_0 \exp\left(-\frac{(t-\tau)^2}{\sigma^2}\right) e^{-i\omega(t-\tau)}, \end{aligned} \tag{B.9}$$

where τ is delay between the pump 1 and pump 2. The intensity $I(\tau)$ detected by the detector is

$$\begin{aligned} I(\tau) &= \int_{-\infty}^{\infty} |E_1(t) + E_2(t)|^2 \, dt = \int_{-\infty}^{\infty} (E_1(t) + E_2(t))(E_1^*(t) + E_2^*(t)) dt \\ &= |E_0|^2 \int_{-\infty}^{\infty} \exp\left(-\frac{2t^2}{\sigma^2}\right) + \exp\left(-\frac{2(t-\tau)^2}{\sigma^2}\right) \\ &\quad + 2\exp\left(-\frac{t^2 + (t-\tau)^2}{\sigma^2}\right) \cos(\omega\tau) dt \\ &= |E_0|^2 \sqrt{2\pi}\sigma \left(1 + \exp\left(-\frac{\tau^2}{2\sigma^2}\right) \cos(\omega\tau)\right). \end{aligned} \tag{B.10}$$

Appendix C
Mathematical Formula

C.1 Gaussian Integral

The Gaussian integral is

$$I(a) = \int_{-\infty}^{\infty} e^{-ax^2} \mathrm{d}x = \sqrt{\frac{\pi}{a}}, \tag{C.1}$$

where Re(a) > 0.

The square of I is calculated as

$$\begin{aligned} I(a)^2 &= \left(\int_{-\infty}^{\infty} e^{-ax^2} \mathrm{d}x\right)^2 = \left(\int_{-\infty}^{\infty} e^{-ax^2} \mathrm{d}x\right)\left(\int_{-\infty}^{\infty} e^{-ay^2} \mathrm{d}y\right) \\ &= \int_{-\infty}^{\infty}\int_{-\infty}^{\infty} e^{-a(x^2+y^2)} \mathrm{d}x\mathrm{d}y. \end{aligned} \tag{C.2}$$

Using polar coordinates ($x = r\cos\theta$ and $y = r\sin\theta$), the integral is

$$\begin{aligned} I(a)^2 &= \int_0^{\infty}\int_0^{2\pi} e^{-ar^2} \mathrm{d}\theta r \mathrm{d}r = 2\pi \int_0^{\infty} e^{-ar^2} r \mathrm{d}r \\ &= 2\pi \left[\frac{e^{-ar^2}}{-2a}\right]_0^{\infty} = \frac{\pi}{a}. \end{aligned} \tag{C.3}$$

Finally, we get $I(a) = \sqrt{\pi/a}$.

C.1.1 Two Complex Parameters' Case

Let us consider the Gaussian integral with two complex parameters (α and β)

K. Nakamura, *Quantum Phononics*, Springer Tracts in Modern Physics 282,
https://doi.org/10.1007/978-3-030-11924-9

$$I(\alpha, \beta) = \int_{-\infty}^{\infty} e^{-\alpha(x+\beta)^2} \mathrm{d}x, \tag{C.4}$$

where $\mathrm{Re}(\alpha) > 0$.[2] The integrand function goes to zero at $x \to \pm\infty$ for the $\mathrm{Re}(\alpha)>0$ condition, and the integral $I(\alpha, \beta)$ converges. The partial derivative of $I(\alpha, \beta)$ by β is

$$\begin{aligned}\frac{\partial I(\alpha, \beta)}{\partial \beta} &= \int_{-\infty}^{\infty} \frac{\partial}{\partial \beta} e^{-\alpha(x+\beta)^2} \mathrm{d}x = \int_{-\infty}^{\infty} -2\alpha(x+\beta) e^{-\alpha(x+\beta)^2} \mathrm{d}x \\ &= \left[e^{-\alpha(x+\beta)^2} \right]_{-\infty}^{\infty} = 0. \end{aligned} \tag{C.5}$$

Then $I(\alpha, \beta)$ is independent on β. The partial derivative of $I(\alpha, \beta)$ by α is

$$\begin{aligned}\frac{\partial I(\alpha, \beta)}{\partial \alpha} &= \int_{-\infty}^{\infty} -(x+\beta)^2 e^{-\alpha(x+\beta)^2} \mathrm{d}x \\ &= \int_{-\infty}^{\infty} \frac{x+\beta}{2\alpha} \left(-2\alpha(x+\beta) e^{-\alpha(x+\beta)^2} \right) \mathrm{d}x \\ &= \left[\frac{x+\beta}{2\alpha} e^{-\alpha(x+\beta)^2} \right]_{-\infty}^{\infty} - \int_{-\infty}^{\infty} \frac{1}{2\alpha} e^{-\alpha(x+\beta)^2} \mathrm{d}x \\ &= -\frac{1}{2\alpha} I(\alpha, \beta). \end{aligned} \tag{C.6}$$

Therefore, we get

$$\begin{aligned} I(\alpha, \beta) &= C \exp\left(-\frac{1}{2} \ln \alpha \right) = C \exp\left(-\frac{1}{2} \ln(|\alpha| + i \arg(\alpha)) \right) \\ &= \frac{C}{\sqrt{|\alpha|}} \exp\left(-\frac{i}{2} \arg(\alpha) \right), \end{aligned} \tag{C.7}$$

where C is a constant, and $\arg(\alpha)$ is the argument of α. If we set $\alpha = a > 0$, $\arg(\mathrm{a}) = 0$ and

$$\begin{aligned} I(a, \beta) &= \frac{C}{\sqrt{a}} \\ &= I(a) = \sqrt{\frac{\pi}{a}}. \end{aligned} \tag{C.8}$$

Then we get $C = \sqrt{\pi}$. Finally, we get

$$I(\alpha, \beta) = \int_{-\infty}^{\infty} e^{-\alpha(x+\beta)^2} \mathrm{d}x = \sqrt{\frac{\pi}{|\alpha|}} \exp\left(-\frac{i}{2} \arg(\alpha) \right). \tag{C.9}$$

---

[2] We use the method described in [5].

Using this formula, we get

$$\int_{-\infty}^{\infty} e^{-a(x+ib)^2} \mathrm{d}x = \sqrt{\frac{\pi}{a}}, \tag{C.10}$$

where a and b are real.

C.1.2 Fourier Transformation

Fourier transformation of the Gaussian function is also important and often used in the text.

$$\begin{aligned}\int_{-\infty}^{\infty} e^{-ax^2} e^{ikx} \mathrm{d}x &= \int_{-\infty}^{\infty} e^{-a(x^2-ikx/a)} \mathrm{d}x \\ &= \int_{-\infty}^{\infty} e^{-a(x-ik/(2a))^2} e^{-k^2/(4a)} \mathrm{d}x \\ &= I(a, -ik/(2a)) e^{-k^2/(4a)} = \sqrt{\frac{\pi}{a}} e^{-k^2/(4a)}. \end{aligned} \tag{C.11}$$

$I(a, -ik/(2a))$ is obtained by using (C.9) and $\arg(a) = 0$.

References

1. Bruesch, P.: Phonons: Theory and Experiments I. Springer Series in Solid-State Sciences, vol. 34. Springer, Berlin (1982)
2. Rulliere, C. (ed.): Femtosecond Laser Pulses, Principles and Experiments, 2nd edn. Springer Science+Business Media Inc, Berlin (2005)
3. Watanabe, S.: Cyotanparusu hassei gijyutu (in Japanese: Generation of ultrashort pulse). Kogaku **24**, 378–383 (1995)
4. Trebino, R.: Frequency-Resolved Optical Grating: The measurement of ultrashort laser pulses. Kluwer Academic Publishers, Norwell (2000)
5. Nomoto: Gauss sekibun no ippannka (in Japanese: General Solution of Gaussian Integral). http://www.eng.niigata-u.ac.jp/~nomoto/15.html

Index

K. Nakamura, *Quantum Phononics*, Springer Tracts in Modern Physics 282,
https://doi.org/10.1007/978-3-030-11924-9

T

U

V

W

Z

GPSR Compliance
The European Union's (EU) General Product Safety Regulation (GPSR) is a set of rules that requires consumer products to be safe and our obligations to ensure this.

If you have any concerns about our products, you can contact us on

ProductSafety@springernature.com

In case Publisher is established outside the EU, the EU authorized representative is:

Springer Nature Customer Service Center GmbH
Europaplatz 3
69115 Heidelberg, Germany

www.ingramcontent.com/pod-product-compliance
Ingram Content Group UK Ltd.
Pitfield, Milton Keynes, MK11 3LW, UK
UKHW021824190726
13853UKWH00003B/1182

* 9 7 8 3 0 3 0 1 1 9 2 6 3 *